Springer Natural Hazards

The Springer Natural Hazards series seeks to publish a broad portfolio of scientific books, aiming at researchers, students, and everyone interested in Natural Hazard research. The series includes peer-reviewed monographs, edited volumes, textbooks, and conference proceedings. It covers all categories of hazards such as atmospheric/climatological/oceanographic hazards, storms, tsunamis, floods, avalanches, landslides, erosion, earthquakes, volcanoes, and welcomes book proposals on topics like risk assessment, risk management, and mitigation of hazards, and related subjects.

More information about this series at http://www.springer.com/series/10179

Tariq S. Durrani · Wei Wang
Sheila M Forbes
Editors

Geological Disaster Monitoring Based on Sensor Networks

Springer

Editors
Tariq S. Durrani
Department of Electronic
and Electrical Engineering
University of Strathclyde
Glasgow, UK

Sheila M Forbes
Department of Electronic
and Electrical Engineering
University of Strathclyde
Glasgow, UK

Wei Wang
College of Electronic
and Communication Engineering
Tianjin Normal University
Tianjin, China

ISSN 2365-0656 ISSN 2365-0664 (electronic)
Springer Natural Hazards
ISBN 978-981-13-0991-5 ISBN 978-981-13-0992-2 (eBook)
https://doi.org/10.1007/978-981-13-0992-2

Library of Congress Control Number: 2018945877

This Springer imprint is published by the registered company Springer Nature Singapore Pte Ltd.
The registered company address is: 152 Beach Road, #21-01/04 Gateway East, Singapore 189721, Singapore

Contents

Introduction

Tariq S. Durrani, Wei Wang and Sheila M Forbes

Abstract Large scale natural disasters cause untold misery and massive damage to life, infrastructure and property. Such disasters, often categorised as geophysical (such as earthquakes, volcanic eruptions, tsunamis, landslides, snowdrifts and avalanches), hydrological (including floods, river and debris overflows), meteorological (hurricane, tropical storms, sandstorms, high winds, heavy rainfall), climatological (such as wild-fires, drought, extreme temperatures), lead to significant loss of life, damage to the living, human displacement and poverty and indeed to devastation of the foundations of cities, towns, villages and the countryside; and the associated damage to the infrastructure of roads, housing, buildings, bridges, communication systems and more. Victims are often trapped in collapsed buildings, without electricity, water or other means of communications. Thus the development and understanding of advanced techniques for disaster relief are of immense current interest, and there is a compelling need for effective disaster prediction, relief, and associated management systems and the development and understanding of advanced techniques for disaster relief are of immense current interest. Requirements for the enhancement of early warning and emergency response systems to geological disasters are of essential importance. To ensure speedy recovery of people and the protection of the national infrastructure threatened by natural disasters, real time detection and data collections are a necessary prerequisite. Threats become even more complex due to the evolution of geological disasters.

Keywords Geological disasters · Disaster monitoring · Networks

---

T. S. Durrani (✉) · S. M. Forbes
University of Strathclyde, Glasgow, Scotland, UK
e-mail: durrani@strath.ac.uk

S. M. Forbes
e-mail: sheila.forbes@strath.ac.uk

W. Wang
Tianjin Normal University, Tianjin, People's Republic of China
e-mail: weiwang@tjnu.edu.cn

T. S. Durrani et al. (eds.), *Geological Disaster Monitoring Based on Sensor Networks*, Springer Natural Hazards, https://doi.org/10.1007/978-981-13-0992-2_1

Large scale natural disasters cause untold misery and massive damage to life, infrastructure and property. Such disasters, often categorised as geophysical (such as earthquakes, volcanic eruptions, tsunamis, landslides, snowdrifts and avalanches), hydrological including floods, river and debris overflows), meteorological (hurricane, tropical storms, sandstorms, high winds, heavy rainfall), climatological (such as wild-fires, drought, extreme temperatures), lead to significant loss of life, damage to the living, human displacement and poverty and indeed to devastation of the foundations of cities, towns, villages and the countryside; and the associated damage to the infrastructure of roads, housing, buildings, bridges, communication systems and more. Victims are often trapped in collapsed buildings, without electricity, water or means of communications. Thus there is a compelling need for effective disaster prediction, relief, and associated management systems, and the development and understanding of advanced techniques for disaster relief are of immense current interest. Requirements for the enhancement of early warning and emergency response systems to geological disasters are of essential importance. To ensure speedy recovery of people and the protection of the national infrastructure threatened by natural disasters, real time detection and data collection is a necessary prerequisite. Threats become even more complex due to the evolution of geological disasters.

Construction of geological condition monitoring sensor networks in areas prone to earthquakes, volcanoes, and landsides would provide information on geographical structural state changes through the real time online analysis of large scale sensor networks data. Such networks would also provide early warning of major geological disasters, reduce casualties and property losses.

To explore these areas of concern and to identify developments of new technologies for monitoring susceptible locations, an International Workshop was held in Harbin, China, from 14 to 17 July 2017, where some of the leading workers in the field presented their latest research findings. The Workshop was funded by the National Natural Science Foundation of China (NSFC) and the British Council under the Newton Researcher Links Programme.

The Workshop brought together some thirty early career researchers from China and the UK, with complementary skills in geosciences, electronics, wireless systems, and sensor networks. The objectives of the Workshop were to increase research capacity, encourage knowledge transfer from cognate areas, explore and identify opportunities for collaborative research, and build international teams for future collaboration in this area of international importance—disaster recovery and mitigation. The aim of the Workshop was to identify emerging areas of research in wireless sensor networks with high impact potential on disaster monitoring; addressing real world situations using advanced sensor and signal processing and communications technologies.

This monograph is the outcome of the proceedings of the Workshop. The Editors have carefully selected the most informative, challenging and original of the presentations, which have been further carefully nurtured by the authors for inclusion in this Monograph.

The Second Chapter by Heiderzadeh and Gusman et al. "Application of Dense Offshore Tsunami Observations from Ocean Bottom Pressure Gauges (OBPGs) for Tsunami Research and Early Warnings", covers a relatively new area in Disaster monitoring—that of Tsunami observations from deep Ocean Bottom Pressure Gauges (OBPGS), which offer new insights into tsunami characteristics. The authors present a procedure for extracting tsunami signals from OBPGs data, and then apply the procedure to two tsunami case studies to verify the efficacy of their approach and the value of the use of OBPG data towards tsunami research and warnings.

The follow-on Third Chapter by Novellino et al. "Remote Sensing for Natural or Man-Made Disasters and Environmental Changes", is concerned with the effective use of satellite remote sensing in supporting disaster management studies in areas affected by natural hazards. The authors contend that a key reason for the adoption of remote sensing is that it is one of the fastest means of acquiring data in a timely and cost effective manner, up to regional-scale during pre-disaster and post-disaster studies. Using three distinctive case studies from (i) the landslide inventory map of St. Lucia island, (ii) tsunami-induced damage along the Sendai coast (Japan) and (iii) the landslide geotechnical characterization in Papanice (Italy), the authors show recent advances in remote sensing, including the use of new spaceborne/airborne sensors and techniques, that offer a 'best practice' environment for the better management of geohazards.

The next two Chapters (Four and Five) "Classification of Post-Earthquake High Resolution Image Using Adaptive Dynamic Region Merging and Gravitational Self-organizing Maps" and "A Survey on the Role of Wireless Sensor Networks and IoT in Disaster Management" are complementary, in that they deal with the acquisition and handling of big data and its analysis, which in one case addresses the issue of disaster monitoring and recovery, and the other studies post-disaster issues confronting the management scenarios. The authors in Chapter Four "Classification of Post-Earthquake High Resolution Image Using Adaptive Dynamic Region Merging and Gravitational Self-organizing Maps" have addressed the important issue of the recognition of regions of similarity and dissimilarity needed to classify areas of common relevance following an earthquake. To this effect the authors describe their work as developing advanced segmentation tools for image processing using adaptive and dynamic region merging and combining these with sophisticated feature extraction techniques that rely on spectral and spatial feature textures to devise gravitational self-organizing maps (gSOM) that offer a novel object-based classification framework. The work is well illustrated by application to data and aerial seismic images from the Wenchuan earth quake of 2008 to demonstrate the effectiveness of the proposed techniques. The methods while conceptually advanced are computationally expensive.

Adeel et al., have conducted a review of emerging technologies for communications that aid disaster management, and have identified a range of instruments applicable to wireless sensor networks, including the use of 4G and 5G systems, and indeed the emerging technologies of the 'Internet of Things (IoT)' and the associated big data technologies. A valuable piece of work in the Chapter includes an evaluation of two major IoT standards; and a potential solution based on Cognitive 5G long range and low power sensor networks.

The Sixth chapter by Ilya Sianko et al., on "Modelling of Earthquake Hazard and Secondary Effects for Loss Assessment in Marmara (Turkey)", reports on the excellent work on 'Earthquake Risk Assessment' that has been conducted at the University of Sheffield. One of the most critical components in seismic risk assessment is the calculation of the hazard, and this Chapter proposes tools for determining earthquake hazards, especially for regions with limited seismo-tectonic information. They have developed a seismic hazard analysis tool, based on probabilistic modelling, which generates synthetic earthquakes using a Monte Carlo approach; and have carried out a case study on the area of Marmara in Turkey to validate the effectiveness of the tool.

In the Seventh Chapter "Unmanned Aerial Vehicles for Disaster Management", Lou et al., present a comprehensive study of the use of Unmanned Aerial Vehicles (UAVs) as an effective strategy for disaster management and response, in practical environments. Basing their work on the premise that UAVs can be easily deployed and can reach inaccessible locations, they make a compelling case for the deployment of UAVs to map out affected areas in a relatively short time; and aiding a swift and efficient response to a disaster by providing accurate hazard maps, in high resolution and in real time, as an effective guide to the rescuer to assess the situation, make relief plans and conduct rescue.

They analyse networked architecture for multiple UAVs, which represents an enhanced and efficient network-assisted disaster management system that involves data collection, victim localisation and rescue optimisation. They introduce a universal networked architecture that integrates WiFi, cellular, self-managed UAV ad hoc and satellite networks, to offer easy and fast-to-deploy, flexible, and inexpensive technology to coordinate the rescue teams in the case of disastrous events and to help the survivors in a timely manner. The authors address the associated design and system performance challenges, and include heuristic algorithms for placing UAV nodes to facilitate reliable communications to disconnected groups.

The Eighth chapter on "Human Detection based on Radar Sensor Network in Natural Disaster" by Wei WANG, exploits the use of Ultra-Wideband (UWB) radar technology as a means of detecting humans in disaster affected regions. The premise is based on the ability of UWB radars to penetrate through walls and thus locate humans. UWB systems have been developed using synthetic aperture radar that offer penetration ability and high resolution imaging of hidden 'targets'; and associated Doppler radar systems identifying the presence of human beings by detecting respiratory—induced Doppler signals and human movements.

This Chapter sets the scene by developing the tools based on fuzzy pattern recognition and genetic algorithms for identifying multiple status of human beings from UWB radar returns, and conducting a comprehensive analysis to justify their use. This work is followed by a report on a detailed experiment using a P410 MRM radar device, to assess its performance under six different scenarios-including the through-wall no person status, normal breathing status of one person, swing arms status of one person, normal breathing status of two persons, walking 2 m away status of two persons and normal breathing status of three persons.

The results have a very important theoretical significance and practical value. The difference between the through-wall slow breathing status of three persons and the swing arms status is small, so there is the possibility of wrong judgment or false alarm. This chapter compares the fuzzy pattern recognition algorithm with current standards, and illustrates that the proposed algorithm is superior to the other three algorithms.

The Ninth Chapter on "Real-time Wind Velocity Monitoring based on Acoustic Tomography" introduces another facet of disaster monitoring—i.e. the monitoring of wind—related disasters, and introduces the development and performance of anemometers based on acoustic tomography. These utilise the dependence of sound speed on wind velocity as a promising remote sensing technique for wind velocity monitoring. The approach offers the advantage of being low cost, easy to implement, and non-invasive, i.e. not affecting the localised field.

Using elegant mathematics, the authors develop the theory for the reconstruction of the acoustic wave fields received on an array of acoustic sensors. On the assumption of linear wave fields, the authors use time of flight measurements along multiple ray paths, to develop a method of tomographic reconstruction of wind velocity fields. Using simulated data for three different velocity fields, the authors evaluate the performance of their reconstruction algorithm and show that acoustic tomography provides quality tomographic images of the velocity fields with good accuracy.

The Tenth Chapter on "Joint Optimization of Resource Allocation with Inter-beam Interference for a Multi-beam Satellite and Terrestrial Communication System" is concerned with the study of satellite and terrestrial communication systems in order to evaluate their performance during emergency scenarios depicted by natural disasters, and in providing satellite mobile services. The key concern is optimal resource allocation to conserve on-board resources and their utilisation. Here algorithms are developed to optimise joint bandwidth and power allocation, while taking into consideration the satellite inter-beam interference, channel condition and delay factors.

The work proposes an energy-efficient scheme, which integrates satellite-terrestrial spectrum sharing; based on three stages—firstly the central terrestrial cell receives an intensive signal offering a high signal-to-noise ratio based on a full frequency reuse scheme; secondly ranking the satellite beam isolation for different frequency bands corresponding to the base station/user location, and thirdly, dividing the highest degree of isolation band and the lowest isolation band into one group; the sub-high degree of isolation band and sub-low isolation band into other group and so on.

The authors show that in comparison with current algorithms that offer separate bandwidth or power optimal allocation, their proposed algorithm allocates resources flexibly according to specific traffic demand and channel condition. The authors propose a system model taking into consideration satellite beams, macro base station, remote radio heads and three layers cover that cause serious inter-layer interference. Based on this model, the authors carry out an interference analysis; and then propose an integrated satellite-terrestrial cognitive spectrum sharing scheme based on an exclusion zone, and an associated energy-efficient spectrum allocation scheme with high inter-cell fairness.

Through detailed simulation using realistic scenarios, the authors compare their results with conventional approaches to prove the effectiveness of their approach. They further take into consideration the power consumption model and then develop the joint resource allocation model and analyse this under several scenarios, such as matrix sparseness, water-filling, complexity and power consumption, and then carry out detailed simulation and performance to show that their algorithm outperforms conventional techniques in terms of energy efficiency and activity ratios.

The Eleventh chapter on "Intelligent Sub-meter Localization Based on OFDM Modulation Signal" is a comprehensive study of systems and techniques to be used to optimise Location Based Services (LBS), with associated localisation and navigation applications. While global navigation satellite systems are effective when LBS are sought outdoor, their performance deteriorates considerably in indoor environments; where accurate localisation is a necessity, as in indoor rescues, location of mines, or even finding items in shopping malls. The main contribution of this Chapter, is the design of a new precise indoor localization system based on the CSI (Channel State Information) which is available in many existing commodity Wi-Fi devices to estimate the AOA (Angle of Arrival) of the multipath signal. To overcome the limitation of the conventional AOA estimation approaches, the proposed method exploits the OFDM (Orthogonal Frequency Division Multiplexing) modulation property to estimate the AOA of the signal using a significantly reduced number of antennas, and with a small modification of hardware. To solve this problem, the authors propose the use of two-dimensional spatial smoothing for the AOA estimation with respect to the multiple correlated signals. The proposed system comprises three steps—CSI-based AOA estimation followed by direct signal path identification, and then Target localisation. This offers sub-meter accuracy. The Chapter includes extensive derivation of the associated algorithms, clear justification of the approaches taken, and detailed simulations carried out along with assessment on experimental set ups to evaluate the performance of the proposed approach and compare it with conventional techniques to show the benefits offered by the authors work. The proposed system can be easily implemented on future 5G networks, and LTE (Long-Term Evolution) which is a standard for high-speed wireless communication for mobile devices and data terminals, with advantages of MIMO antennas.

The Twelfth Chapter give an overview of the broad conclusions of the work reported in the earlier chapters, and offers some final comments and observations by the Editors.The Editors would like to thank the Chapter authors and their co-authors for their cooperation, their hard work, and for contributing their insight and experiences to this monograph; and to all the attendees of the Workshop on 'GEOLOGICAL DISASTER MONITORING BASED ON SENSOR NETWORKS held in Harbin in July 2017; and indeed to the sponsors—The National Natural Science Foundation of China and the British Council Researcher Links Programme supported by the Newton Fund.

Application of Dense Offshore Tsunami Observations from Ocean Bottom Pressure Gauges (OBPGs) for Tsunami Research and Early Warnings

Mohammad Heidarzadeh and Aditya R. Gusman

Abstract We introduce a new data source of dense deep-ocean tsunami records from Ocean Bottom Pressure Gauges (OBPGs) which are attached to Ocean Bottom Seismometers (OBS) and apply them for far-field and near-field tsunami warnings. Tsunami observations from OBPGs are new sources of deep-ocean tsunami observations which, for the first time, provide dense tsunami data with spacing intervals in the range of 10–50 km. Such dense data are of importance for tsunami research and warnings and are capable of providing new insights into tsunami characteristics. Here, we present a standard procedure for the processing of the OBPG data and extraction of tsunami signals out of these high-frequency data. Then, the procedure is applied to two tsunamis of 15 July 2009 Mw 7.8 Dusky Sound (offshore New Zealand) and 28 October 2012 Mw 7.8 Haida Gwaii (offshore Canada). We successfully extracted 30 and 57 OBPG data for the two aforesaid tsunamis, respectively. Numerical modeling of tsunami was performed for both tsunamis in order to compare the modeling results with observation and to use the modeling results for the calibration of some of the OBPG data. We successfully employed the OBPG data of the 2012 Haida Gwaii tsunami for tsunami forecast by applying a data assimilation technique. Our results, including two case studies, demonstrate the high potential of OBPG data for contribution to tsunami research and warnings. The procedure developed in this study can be readily applied for the extraction of tsunami signals from OBPG data.

Keywords Tsunami · Ocean Bottom Pressure Gauge · Ocean Bottom Seismometer · Tsunami warning system · Numerical simulation 2009 Dusky Sound earthquake

M. Heidarzadeh (✉)
Department of Civil and Environmental Engineering, Brunel University London, Uxbridge UB8 3PH, UK
e-mail: mohammad.heidarzadeh@brunel.ac.uk

A. R. Gusman
GNS Science, Lower Hutt, New Zealand
e-mail: a.gusman@gns.cri.nz

T. S. Durrani et al. (eds.), *Geological Disaster Monitoring Based on Sensor Networks*, Springer Natural Hazards, https://doi.org/10.1007/978-981-13-0992-2_2

1 Introduction and Background

Tsunami science, in general, is younger than earthquake; mainly because the available observations for tsunamis are less than those for earthquakes. Lack of enough observations has been a main barrier to the development of tsunami science [19]. Tsunami observations are made usually by coastal tide gauges (e.g. [9, 10] and offshore gauges in the form of Deep-ocean Assessment and Reporting of Tsunamis (DART) [2, 3, 8] as well as offshore cabled tsunami gauges such as the Canadian North–East Pacific Underwater Networked Experiments (NEPTUNE) (Rabinovich and Eble [16]. However, most of the tsunami observations have been from tide gauges until 1990s when DARTs were born. Deep-ocean records of tsunamis are free from coastal effects such as harbor resonance [7], nonlinear effect (e.g. [4], and coastal refractions and scattering [11]. Hence, deep-ocean tsunami observations provide refined information about tsunami characteristics [10]. Observations from DARTs are significantly important for tsunami research and warnings and have provided the opportunity to study ocean-wide propagation of tsunamis and to develop a tsunami warning system in the Pacific Ocean [20]. The total number of DARTs installed in the Pacific, Atlantic and Indian Oceans is ~60. Although installation and maintenance of this number of DARTs is a major progress worldwide in tsunami research and has been very costly (installation of each DART approximately costs US$250k), it is not enough to provide high spatial resolution of trans-Pacific tsunamis. The distances between neighboring DARTs are in the range 400–4000 km. Given a wavelength of upto ~500 km for tsunami waves in deep-ocean, it is clear that DART records are very sparse to capture a full tsunami wavelength. In fact, the available deep-ocean measurements of tsunamis through DARTs are limited and sparse. Therefore, it is necessary to look for alternate complementary sources of deep-ocean tsunami measurements.

In past few years, Ocean Bottom Pressure Gauges (OBPG) were added to Ocean Bottom Seismometers (OBS); thus OBSs have been able to record tsunami waves in addition to seismic waves. Because OBSs are deployed in a dense array (upto around 100 instruments) with spacing of 10–50 km, the tsunami records by OBPGs have high spatial resolution. Figures 1 shows dense OBSs which have been deployed in past few years in world's oceans. Some of these OBS systems have been equipped with OBPGs which enabled them to record the trans-oceanic tsunamis (Fig. 1). According to Fig. 1, among the recorded tsunami events by OBPGs are the 2009 Dusky Sound (offshore New Zealand), the 2011 Japan and the 2012 Haida Gwaii (offshore Canada) events.

OBPGs are different from DARTs in several ways: (1) OBSs are usually deployed for few-year campaigns and thus are not permanent stations whereas DARTs are permanent, (2) OBSs store the sea-level data in their hard disks which can be accessed usually at the end of the campaigns or at certain intervals while DARTs provide real-time data through satellite connections, (3) the OBS data have

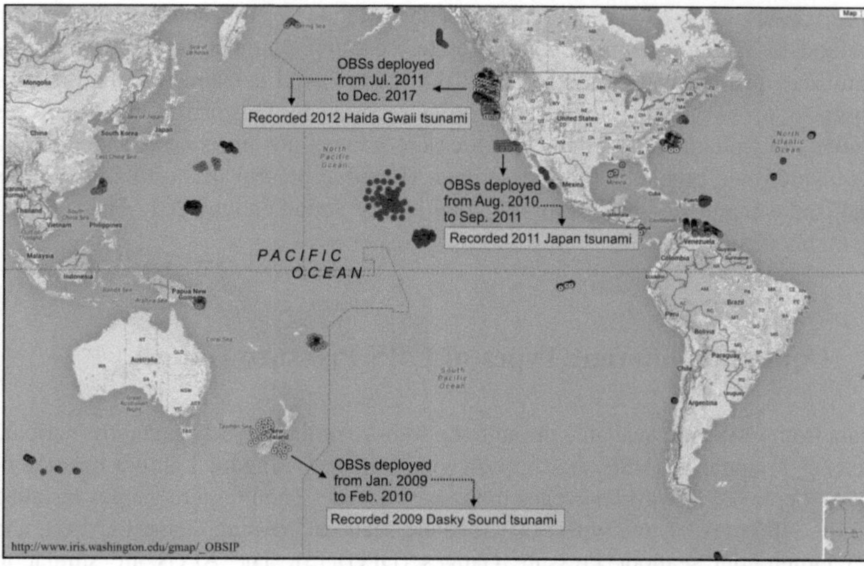

Fig. 1 Locations of OBS campaigns deployed in world's oceans which record both seismic and tsunami waves through OBPGs (original figure from: http://www.iris.washington.edu/gmap/_OBSIP). The three tsunamis of 2009 Dusky Sound, 2011 Japan and 2012 Haida Gwaii were recorded by the OBS systems through their OBPGs

high sampling rates of 10–50 samples per second while DARTs record the tsunami waves with a rate of 1 record per 15 s at best, and (4) OBSs are deployed in large numbers (from ~50 to ~100) with spacing in the range 10–50 km (Fig. 1) whereas DARTs are limited in number (total number of DARTs is ~60 world-wide) and are spaced from ~400 to ~4000 km.

Dense OBPG observations are helpful for tsunami research and warnings. While temporal variations of tsunamis are well known by having a large number of time series of tsunamis, little is known about spatial variations of tsunamis because tsunamis have large wavelengths (i.e. hundreds of kilometers) and dense array of tsunamis have not been available so far. Therefore, it has been impossible to provide several measurements of tsunamis per wavelength as they travel across the world's oceans. Data from dense array of OBS pressure gauges provide several measurements per tsunami wavelength; thus can help to study spatial distribution of tsunamis. In addition, dense array of tsunamis provides new opportunities for tsunami warnings by new methods such as warnings based on direct sea-surface measurements (without knowledge about earthquake source), and successive data assimilations (e.g. [5, 15]). Application of both of the aforesaid methods has not been possible for tsunami research so far because such methods require dense observations; i.e. several measurements per tsunami wavelength which means observations at 5–20 km intervals. Maeda et al. [15] proposed an assimilation method for tsunami warning which was tested using synthetic data. The real tsunami data provided by OBSs for the 2012 Haida Gwaii tsunami was the first real application of data assimilation method as reported by Gusman et al. [5]. In this study, the tsunami data from OBS pressure gauges are introduced and the data acquisition and preparation are described. Here, we present the results of OBPGs data and tsunami simulations for the 2009 Dusky Sound and the 2012 Haida Gwaii tsunamis.

2 Data and Different Types of OBS Pressure Gauges

Data from OBSs are available through the website of the project funded by National Science Foundation (NSF) at: <http://www.obsip.org/>. Figure 1 shows location of OBSs deployed in world's oceans in the past decade. The pressure gauges installed on the OBSs are of two types: (1) Absolute seafloor Pressure Gauges (APG), and (2) Differential seafloor Pressure Gauges (DPG) [5]. The APGs are similar to DARTs and give absolute values of pressure above the instrument. DPGs measure the difference between water pressure above the instrument and the oil pressure within the instrument. Hence, the wave amplitudes obtained from DPGs need calibration. Examples of instrument response for the APGs and DPGs at different frequencies are given in Fig. 2. It can be seen that APGs' response is constant at the tsunami period band (2 min < period < 100 min) (Fig. 2a) while the response decreases with increase of period for DPGs (Fig. 2b). In other words, the tsunami amplitudes recorded by DPGs are relative values and do not represent the real tsunami amplitudes while their periods are correct. Therefore, amplitudes of DPGs need correction.

In the past decade, few tsunamis have been recorded by OBS pressure gauges among which are the 2009 Dusky Sound tsunami (New Zealand) (Fig. 3), the 2011 Japan tsunami (Fig. 4), and the 2012 Haida Gwaii tsunami (Fig. 5). Figure 6 presents examples of DART, APG and DPG records of the 2012 Haida Gwaii tsunami and comparisons with simulated waveforms. As shown in Fig. 6, the amplitudes of the waves recorded by DPGs are larger than those recorded by neighboring DARTs and APGs. This is because of the differential nature of the pressures recorded by the DPG instruments and thus the records need to be corrected. However, the periods of the waves recorded by DPGs are the same as those recorded by APGs and DARTs. Besides the aforesaid three events, other tsunamis also were recorded by the OBS arrays such as the 1 April 2014 Iquique (Chile) tsunami.

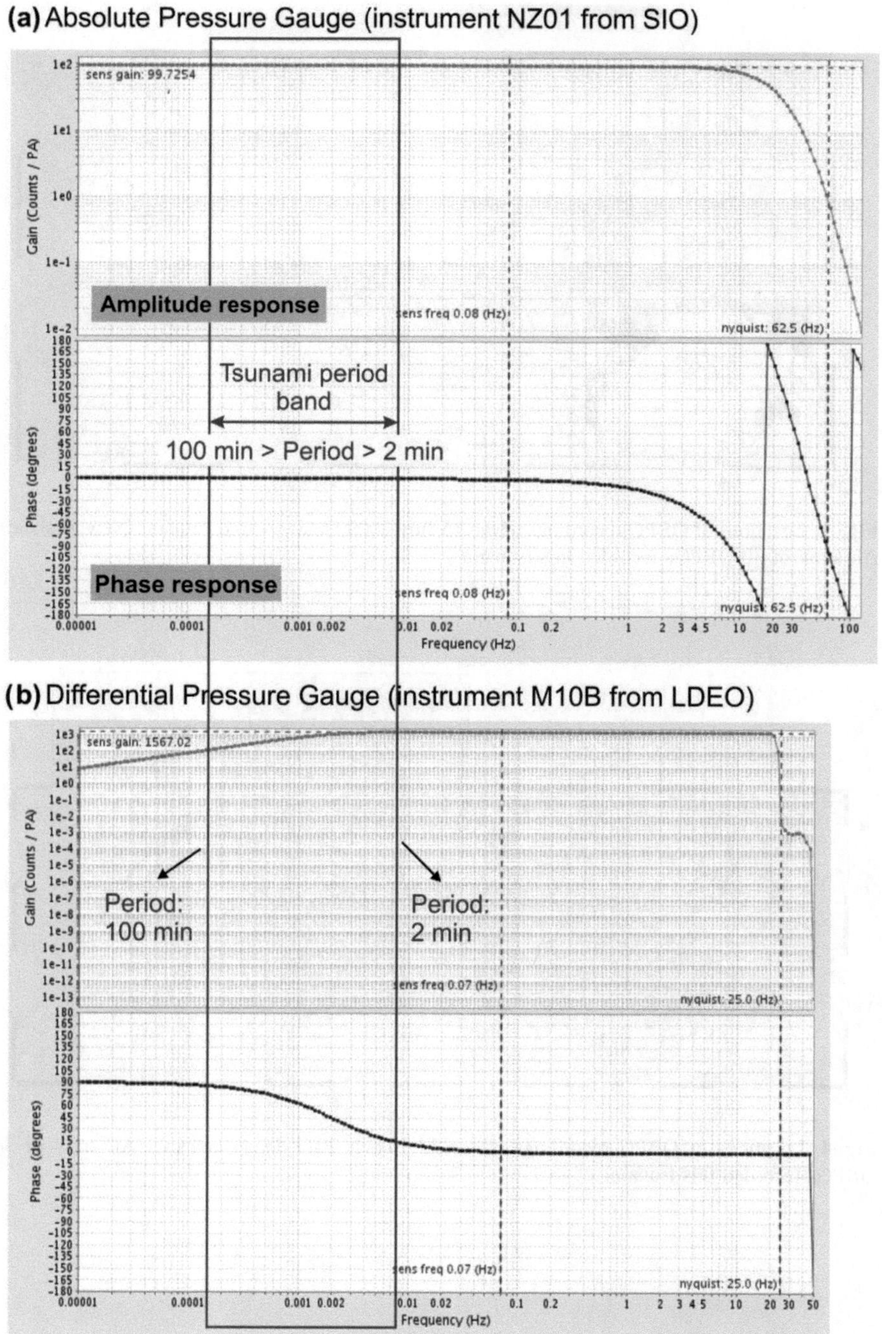

Fig. 2 Sample instrument response for the amplitudes and phases gains at different frequencies for an APG (**a**) and a DPG instrument (**b**). SIO and LDEO stand for Scripps Institution of Oceanography and Lamont-Doherty Earth Observatory, respectively. Data from: Incorporated Research Institutions for Seismology Data Management Center (http://ds.iris.edu/mda/_OBSIP)

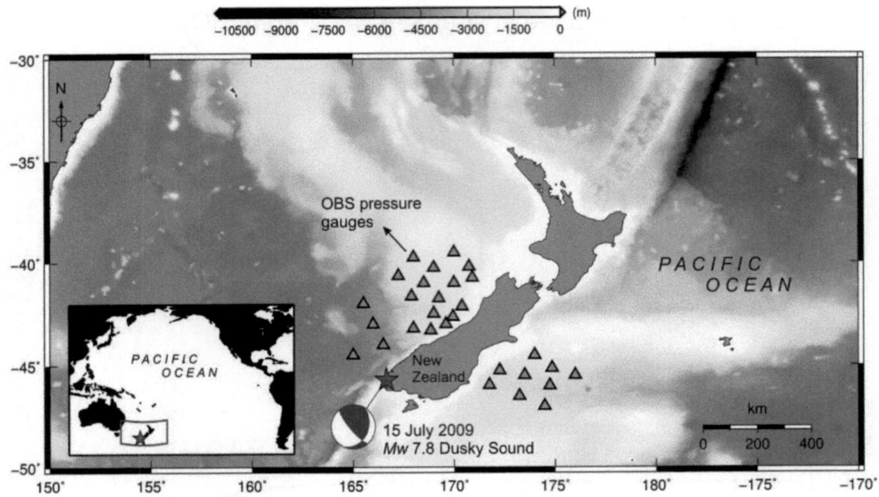

Fig. 3 Locations of OBPG recordings of the 15 July 2009 Dusky Sound tsunami (New Zealand). An array of 30 OBPGs recorded this tsunami

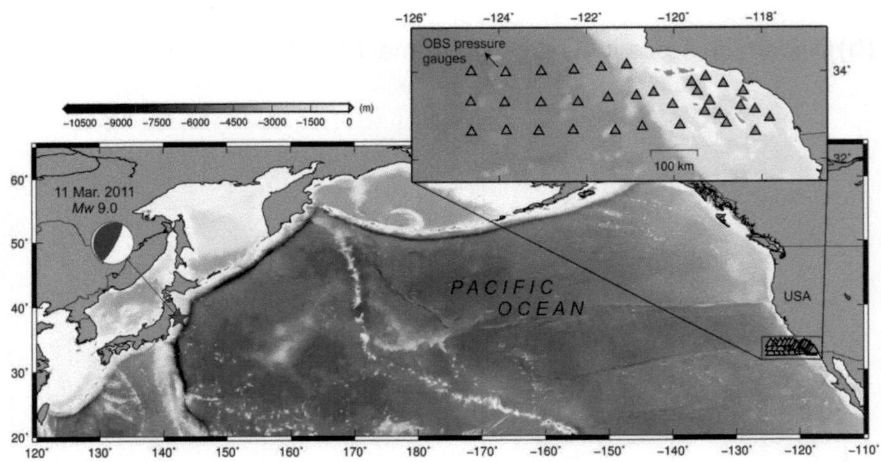

Fig. 4 Locations of OBPG recordings of the 11 March 2011 Japan tsunami. An array of 34 OBPGs recorded this tsunami

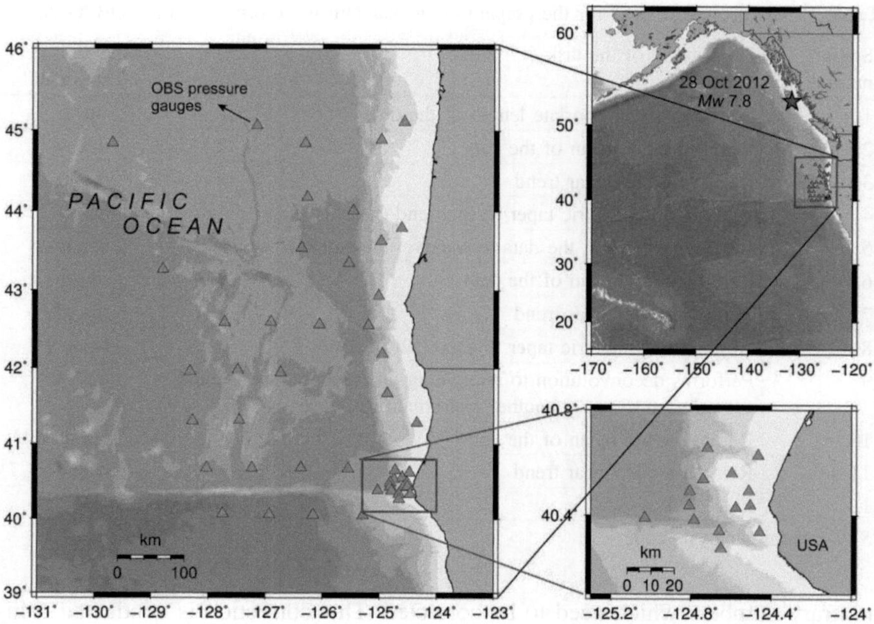

Fig. 5 Locations of OBPG recordings of the 28 November 2012 Haida Gwaii tsunami. An array of 68 OBPGs recorded this tsunami

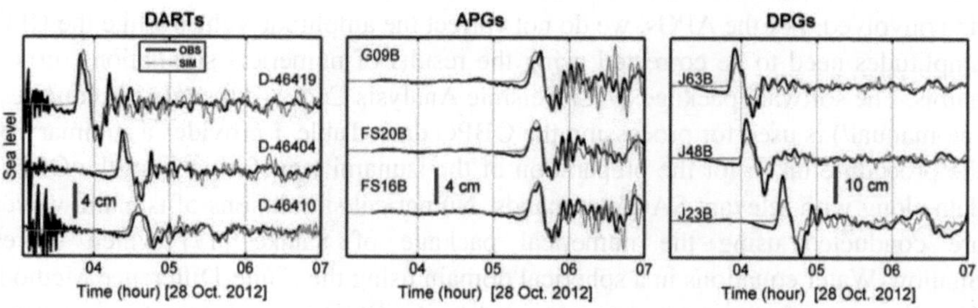

Fig. 6 Examples of DART (left), APG (middle) and DPG (right) records of the 2012 Haida Gwaii tsunami. Black and red waveforms are observed and simulated waveforms, respectively. The observed waveforms from DPGs are noticeably larger than those from DARTs and APGs showing that DPGs need correction (Color figure online)

3 Methodology

Unlike Tide Gauge (TG) or DART data, the process of OBPG data is more complicated. Usually, the amplitude values for the TG and DART data are the absolute real-world values. Therefore, a simple high-pass filter will yield the tsunami signal for the TG and DART data. For two types of OBPG data, the APGs give the absolute values of wave amplitude (same as TG and DARTs) while DPGs give

Table 1 The procedure used for the preparation of tsunami waveforms from the OBPG data

Step number	Description of the task	SAC[a] command
1	Selecting an appropriate length of the data	cut
2	Removing the mean of the data	rmean
3	Removing the linear trend	rtrend
4	Appling a symmetric taper to each end of data	taper
5	Band pass filtering the data to remove non-tsunami signals	bandpass
6	Removing the mean of the data	rmean
7	Removing the linear trend	rtrend
8	Appling a symmetric taper to each end of data	taper
9	Performs deconvolution to remove an instrument response and convolution to apply another instrument response	transfer
10	Removing the mean of the data	rmean
11	Removing the linear trend	rtrend
12	End	

[a]*SAC* Seismic analysis code

arbitrary numbers which need to be corrected. This correction is conducted using the results of tsunami simulations [5].

To extract the tsunami signals from OBPGs, we first resample the high-frequency date (frequency of 40 or 50 Hz) to a low-frequency data (frequency of 0.0167 Hz), then we band-pass filter the original records; finally the instrument responses are de-convolved. For the APGs, we do not correct the amplitude values while the DPG amplitudes need to be corrected using the results of numerical simulations of tsunamis. The software package SAC (Seismic Analysis Code) (https://ds.iris.edu/files/sac-manual/) is used for processing the OBPG data. Table 1 provides a summary of the procedure taken for the preparation of the tsunami waveforms from the OBPG data along with relevant SAC commands. Numerical simulations of tsunami waves are conducted using the numerical package of Satake [17] which solves Shallow-Water equations in a spherical domain using the Finite-Difference Method. The 30 arc-sec bathymetry data provided by GEBCO is used here for numerical modeling of tsunami [21]. The tsunami source models used for the simulations of the events are based on the model by Gusman et al. [6] for the 2012 Haida Gwaii event (Mw 7.8) and that of Beavan et al. [1] for the 2009 Dusky Sound event (Mw 7.8).

4 Case Study One: The 2012 Haida Gwaii Tsunami, Offshore Canada

On 28 October 2012, 03:04:09 UTC, an earthquake with Mw 7.8, which is known as the 2012 Haida Gwaii earthquake, occurred offshore British Columbia, Canada. The earthquake was initiated at 52.622°N, 132.103°W, at the depth of 14 km [13],

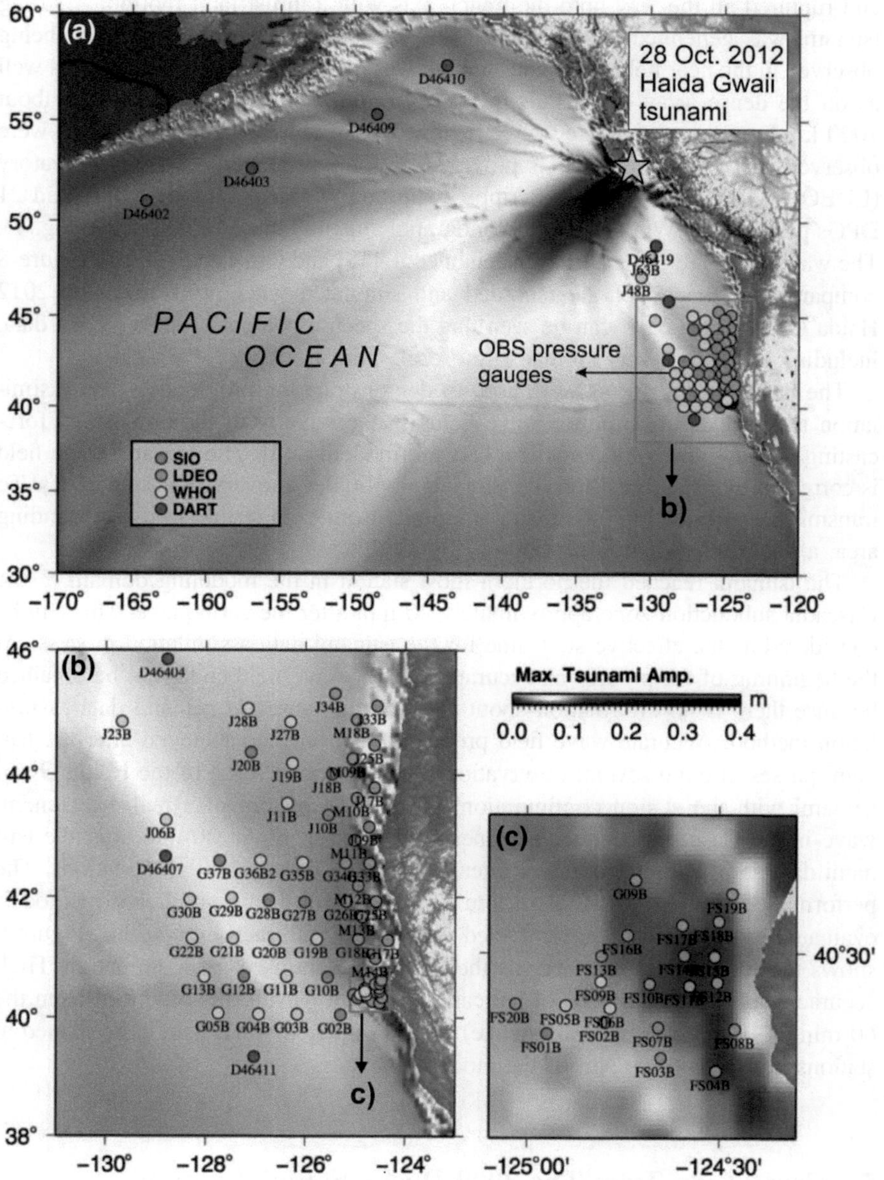

Fig. 7 The maximum simulated tsunami amplitudes due to the 28 November 2012 Haida Gwaii tsunami and locations of DARTs and OBSs. The OBSs are shown by green (Scripps Institution of Oceanography, SIO), brown (Lamont Doherty Earth Observatory, LDEO) and yellow (Woods Hole Oceanographic Institution, WHOI) circles. Modified from Sheehan et al. [18]. An array of more than 50 OBSs recorded this tsunami (Color figure online)

and ruptured all the way upto the trench axis with a thrust fault motion. A strong tsunami was generated by the earthquake with maximum run-up of 13 m being observed in the near field [14]. The tsunami was recorded on DART stations as well as on the dense array of OBPGs in the Cascadia subduction zone located about 1000 km from the earthquake source region. A total of 57 tsunami waveforms were observed at 8 DARTs, 19 APGs provided by Lamont Doherty Earth Observatory (LDEO), 9 DPGs provided by Scripps Institution of Oceanography (SIO), and 21 DPGs provided by Woods Hole Oceanographic Institution (WHOI) [5, 18] (Fig. 7). The waveforms are presented in Sheehan et al. [18] and Gusman et al. [5]. Figure 8 compares the spectra of the recorded and simulated waveforms from the 2012 Haida Gwaii tsunami. It can be seen that the spectral content of all recorded data, including DPGs, are very similar to those of simulations.

The tsunami waveforms were used to demonstrate the progressive data assimilation method [15] to produce wave fields in the vicinity of the array, then forecasting of wave fields by numerical forward modeling [5]. The tsunami wave field is corrected by using the observed tsunami amplitudes at every time step of 1 s. To transmit the information of tsunami amplitude from each station to its surrounding area, a linear interpolation method [12] is used.

The tsunami reached the northern most station in the modeling domain of the Cascadia subduction zone approximately 70 min after the earthquake. This can be considered as the effective start time for the tsunami data assimilation process. At the beginning of the process an accurate tsunami wave field could not be obtained because there is no information about the tsunami source in tsunami data assimilation method. Accurate wave field prediction can only be achieved after the tsunami passes through several observation stations. For the case of the Haida Gwaii tsunami with the station configuration, the general pattern of a realistic tsunami wave in the Cascadia subduction zones begins to emerge at 30 min after the tsunami data assimilation process or after the tsunami passes through 5 stations. The performance of the forecast algorithm using tsunami data assimilation method is evaluated by comparing the forecasted waveforms with the observations. Figure 9 shows the forecast accuracy versus the length of data used for assimilation. High accuracies of more than 80% of forecasted tsunami waveforms produced from the 60 min (130 min after the earthquake) data-assimilated wave field are obtained at stations in the southern part of the modeling area.

5 Case Study Two: The 2009 Dusky Sound Tsunami, Offshore New Zealand

An earthquake with moment magnitude (Mw) of 7.8 occurred in Dusky Sound, New Zealand on 15 July 2009 (see Fig. 10 for epicenter). According to the United States Geological Survey (USGS), the earthquake origin time was 09:22:33 UTC on 15 July 2009, located at 45.722°S 166.64°E and at the depth of 35 km (Fig. 10).

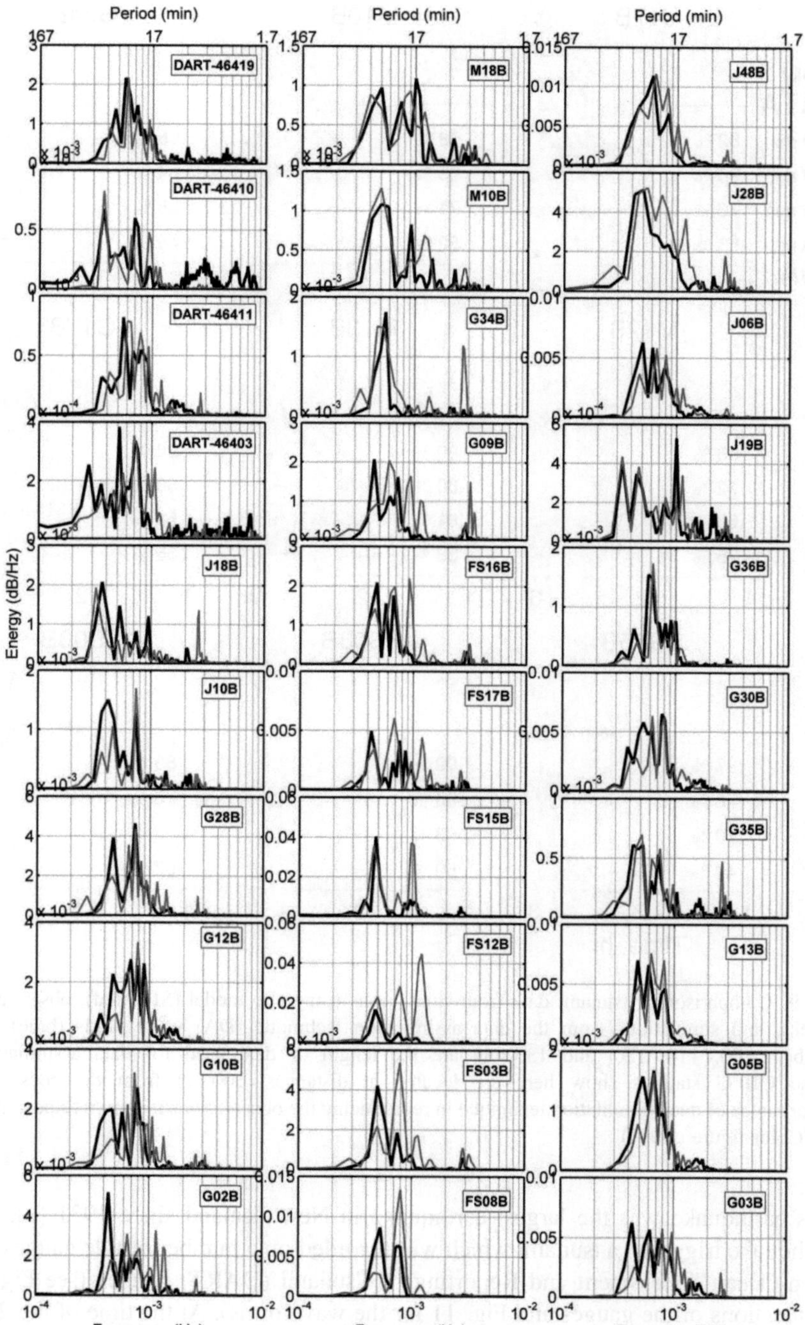

Fig. 8 Comparison of the spectra of the recorded and simulated waveforms from the 2012 Haida Gwaii tsunami

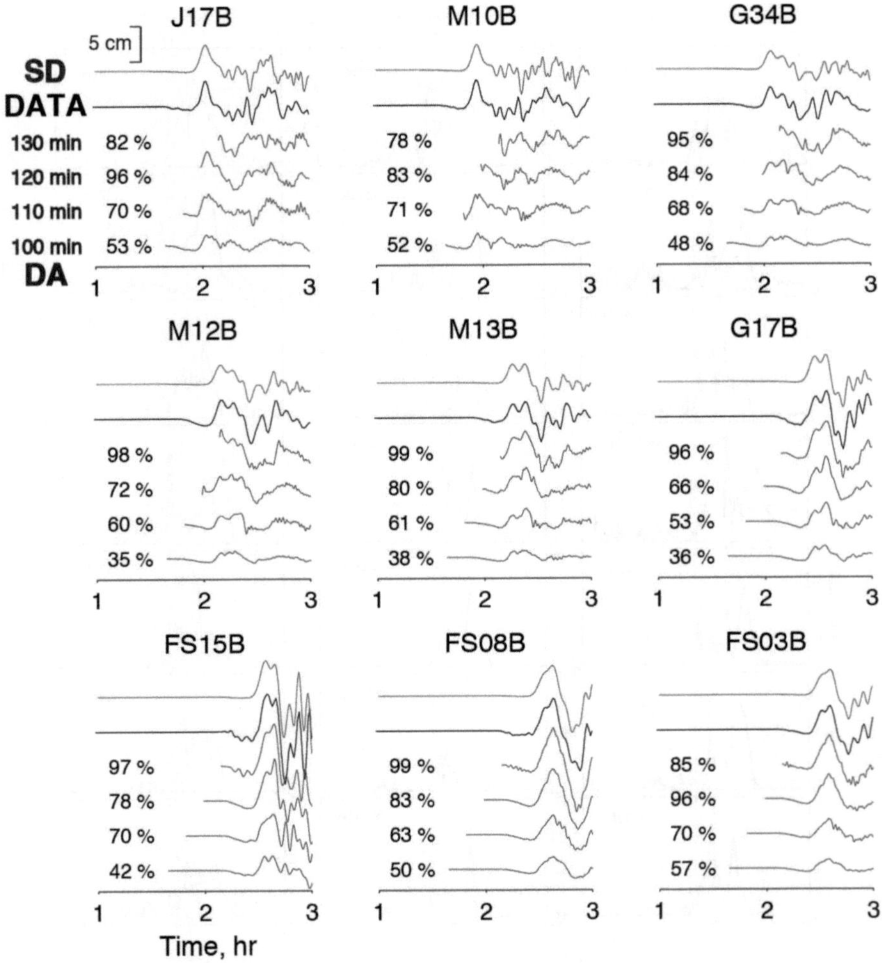

Fig. 9 Comparison of tsunami data from simulations using slip model (SD) (red), observations (black), and simulations from the data assimilation technique (DA) wave fields (blue). The numbers 100, 110, 120, and 130 min are the length of data used for data assimilations. These OBPG stations show here are located at distances <100 km from the coast. The performance of data assimilation technique in reproducing the observations is shown as percentage [5] (Color figure online)

This earthquake was the largest earthquake in New Zealand since 1931 [1]. The earthquake triggered a tsunami which was recorded on a number of tide gauges and Deep-Ocean Assessment and Reporting of Tsunami (DART) gauges (see Fig. 10 for locations of the gauges and Fig. 11 for the waveforms). At the time of the 2009

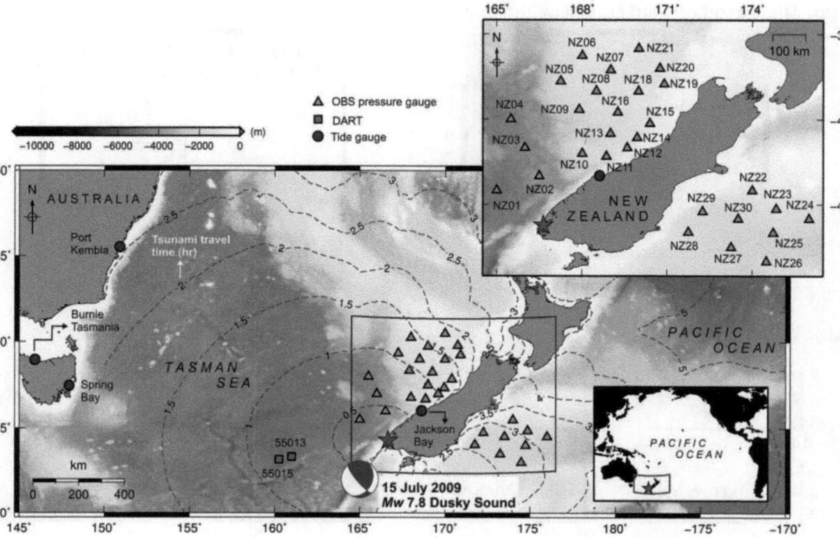

Fig. 10 Epicentral area and location of various sea level gauges used in this study including OBPGs, TGs and DARTs. The red start shows the earthquake epicenter. Dashed contours are tsunami travel times in hours (Color figure online)

earthquake and tsunami, a campaign of OBSs was in operation in the same region (Fig. 10). These OBSs also recorded the tsunami as they were equipped with OBPGs. All of the OBPGs are of the DPG type which means the pressure values are not the absolute values. Therefore, the amplitude values were corrected using the results of tsunami simulations (Fig. 11).

While tsunami signals were fully hidden in high-frequency recordings of the OBPGs, we were able to successfully extract the tsunami signals by applying re-sampling, filtering, and de-convolving the DPG instrument response (the procedure presented in Table 1). In our processed OBPG tsunami data (black lines in Fig. 11), the tsunami arrival times were clear and the signals had periods in the range of 10–20 min which is the expected period range for a tsunami generated by a Mw 7.8 earthquake. Numerical modeling of tsunami was conducted by using the tsunami source proposed by Beavan et al. [1] (Fig. 11a). Simulations were able to fairly reproduce the observations from OBPG, DART and tide gauge stations. However, the amplitudes of the OBPG-DPG data were larger than the simulations; therefore, we corrected the OBPG-DPG amplitudes by applying arbitrary ratios in order to match them with the maximum amplitudes from tsunami simulations for each instrument. Based on Fig. 11, the match for DART and tide gauge records was better than that for OBPGs.

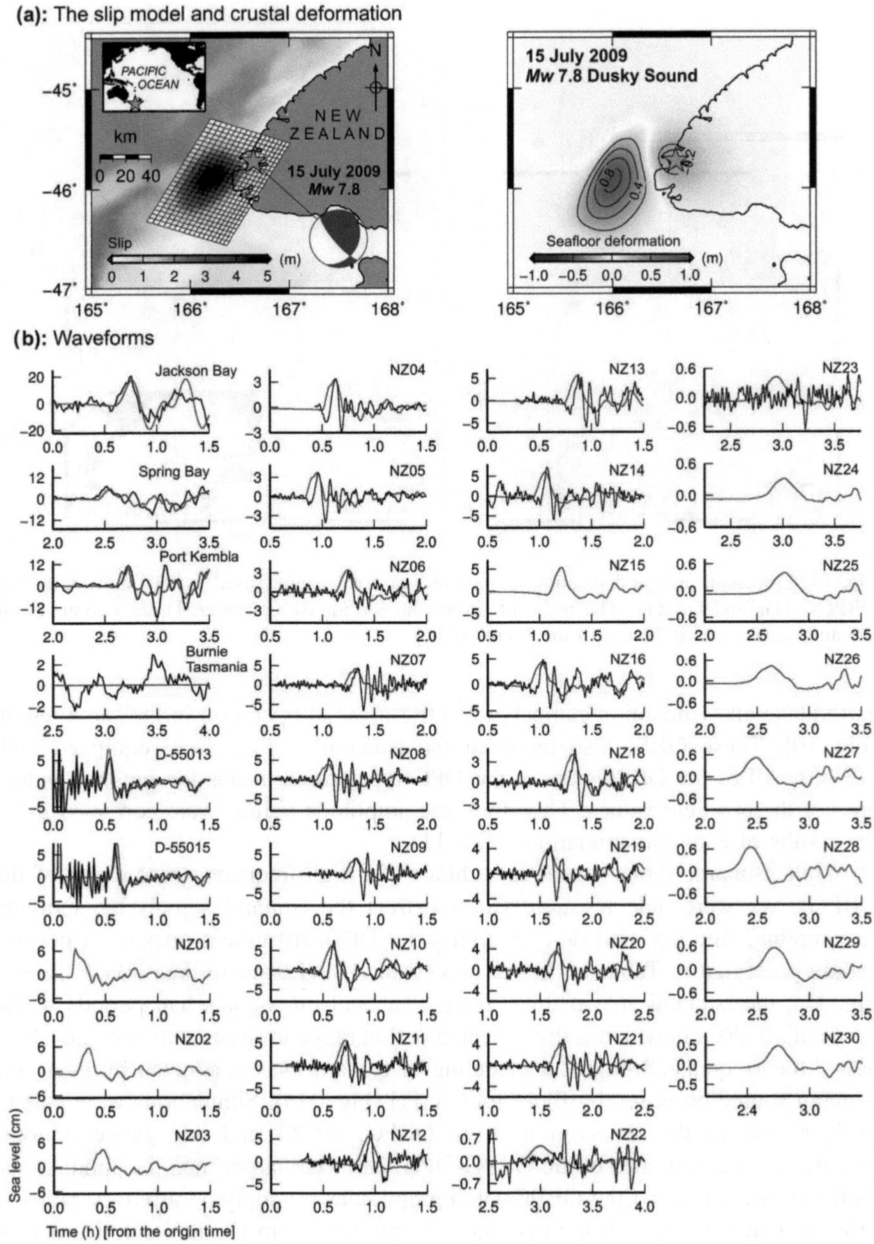

Fig. 11 **a** Source model of the 2009 earthquake according to the model published by Beavan et al. [1]. **b** Comparison of observed (black) and simulated (red) tsunami waveforms for the 2009 Dusky Sound tsunami. The locations of the gauges are shown in Fig. 10. For OBS gauges NZ-15, and from NZ-24 to NZ-30, the tsunami signals are not clear and are hidden within the noise level (Color figure online)

6 Conclusions

We introduced a new source of dense offshore tsunami observations from Ocean Bottom Pressure Gauges (OBPGs) which are attached to Ocean Bottom Seismometers (OBSs). Until recently (i.e. around 2015), offshore deep-ocean tsunami observations were made through DARTs (Deep-ocean Assessment and Reporting of Tsunamis). However, OBPG observations have two main advantages over DARTs namely: (1) they come with large numbers (upto ~ 100) and dense distribution with spacing of 10–50 km versus 200–4000 km of DARTs, and (2) they have high frequency with sampling rates of 40–100 Hz versus that of 0.016 Hz for DARTs. The data processing and preparations are more complicated for OBPGs than DARTs. We presented a standard procedure and the sequence of tasks that needs to be taken for the processing of the OBPG data and extraction of the tsunami signals. The procedure is then applied to the two tsunamis of 2009 Dusky Sound (offshore New Zealand) and the 2012 Haida Gwaii (offshore Canada). Our results showed that the standard procedure used for the extraction of the OBPG data was successful in revealing tsunami signals in both cases. The OBPG instruments for these two events were either Differential seafloor Pressure Gauges (DPGs) or Absolute seafloor Pressure Gauges (APGs). The amplitudes from APGs are real values while those from DPGs are relative values and need correction. For the cases of the DPG data, we corrected the amplitudes of the observations signals using the results of tsunami simulations. The OBPG data for the 2012 Haida Gwaii event were successfully applied for tsunami forecast using the data assimilation technique.

Acknowledgements We acknowledge NOAA (National Oceanic and Atmospheric Administration of the US) for providing the DART data (http://www.ndbc.noaa.gov/dart.shtml), the IOC (Intergovernmental Oceanographic Commission) for the tide gauge records (http://www.ioc-sealevelmonitoring.org/) and the Incorporated Research Institutions for Seismology Data Management Center for the OBPG records (http://ds.iris.edu/mda/_OBSIP). Authors would like to thank Kenji Satake (The University of Tokyo, Japan), Tomohiro Takagawa (Port and Airport Research Institute, Japan), Shingo Watada (The University of Tokyo, Japan) and Anne Sheehan (University of Colorado, US) for their collaboration on the analysis of the OPBG records. Parts of this study were previously presented at the AGU (American Geophysical Union) fall meeting in San Francisco (US) in December 2016. The lead author (MH) was funded by the Brunel University London through the Brunel Research Initiative and Enterprise Fund 2017/18 (BUL BRIEF).

References

1. Beavan J, Samsonov S, Denys P, Sutherland R, Palmer N, Denham M (2010) Oblique slip on the Puysegur subduction interface in the 2009 July MW 7.8 Dusky Sound earthquake from GPS and InSAR observations: implications for the tectonics of southwestern New Zealand. Geophys J Int 183(3):1265–1286
2. Geist EL, Titov VV, Synolakis CE (2006) Tsunami: wave of change. Sci Am 294(1):56–63

3. Gonzalez FI, Milburn HM, Bernard EN, Newman JC (1998) Deep-ocean assessment and reporting of tsunamis (DART®): brief overview and status report. In: Proceedings of the international workshop on tsunami disaster mitigation, Tokyo, Japan, 19–22 January 1998
4. Gusman AR, Murotani S, Satake K, Heidarzadeh M, Gunawan E, Watada S, Schurr B (2015) Fault slip distribution of the 2014 Iquique, Chile, earthquake estimated from ocean-wide tsunami waveforms and GPS data. Geophys Res Lett 42:1053–1060
5. Gusman AR, Sheehan A, Satake K, Heidarzadeh M, Mulia IE, Maeda E (2016a) Tsunami data assimilation of Cascadia seafloor pressure gauge records from the 2012 Haida Gwaii earthquake. Geophys Res Lett 43(9):4189–4196
6. Gusman A, Mulia IE, Satake K, Watada S, Heidarzadeh M, Sheehan AF (2016b) Estimate of tsunami source using optimized unit sources and including dispersion effects during tsunami propagation: the 2012 Haida Gwaii earthquake. Geophys Res Lett 43(18):9819–9828
7. Heidarzadeh M, Satake K (2013) The 21 May 2003 Tsunami in the Western Mediterranean sea: statistical and wavelet analyses. Pure Appl Geophys 170(9):1449–1462
8. Heidarzadeh M, Satake K (2013) Waveform and spectral analyses of the 2011 Japan tsunami records on tide gauge and DART stations across the Pacific Ocean. Pure Appl Geophys 170 (6):1275–1293
9. Heidarzadeh M, Satake K (2014) Excitation of basin-wide modes of the Pacific Ocean following the March 2011 Tohoku Tsunami. Pure Appl Geophys 171(12):3405–3419
10. Heidarzadeh M, Satake K, Murotani S, Gusman AR, Watada S (2015) Deep-water characteristics of the Trans-Pacific Tsunami from the 1 April 2014 M w 8.2 Iquique, Chile Earthquake. Pure Appl Geophys 172(3–4):719–730
11. Heidarzadeh M, Harada T, Satake K, Ishibe T, Gusman A (2016) Comparative study of two tsunamigenic earthquakes in the Solomon Islands: 2015 Mw 7.0 normal-fault and 2013 Santa Cruz Mw 8.0 megathrust earthquakes. Geophys Res Lett 43(9):4340–4349
12. Kalnay E (2003) Atmospheric modeling, data assimilation, and predictability. Cambridge University Press, Cambridge, UK
13. Kao H, Shan SJ, Farahbod AM (2015) Source characteristics of the 2012 Haida Gwaii earthquake sequence. Bull Seismol Soc Am 105(2B):1206–1218
14. Leonard LJ, Bednarski JM (2014) Field survey following the 28 October 2012 Haida Gwaii tsunami. Pure Appl Geophys 171(12):3467–3482
15. Maeda T, Obara K, Shinohara M, Kanazawa T, Uehira K (2015) Successive estimation of a tsunami wavefield without earthquake source data: a data assimilation approach toward real-time tsunami forecasting. Geophys Res Lett 42(19):7923–7932
16. Rabinovich AB, Eblé MC (2015) Deep-ocean measurements of tsunami waves. Pure Appl Geophys 172:3281–3312
17. Satake K (1995) Linear and nonlinear computations of the 1992 Nicaragua earthquake tsunami. Pure Appl Geophys 144:455–470
18. Sheehan AF, Gusman AR, Heidarzadeh M, Satake K (2015) Array observations of the 2012 Haida Gwaii tsunami using Cascadia Initiative absolute and differential seafloor pressure gauges. Seismol Res Lett 86(5):1278–1286
19. Synolakis CE, Bernard EN (2006) Tsunami science before and beyond Boxing Day 2004. Philos Trans R Soc Lond A 364(1845):2231–2265
20. Titov VV, Gonzalez FI, Bernard EN, Eble MC, Mofjeld HO, Newman JC, Venturato AJ (2005) Real-time tsunami forecasting: challenges and solutions. In: Developing tsunami-resilient communities. Springer, Netherlands, pp 41–58
21. Weatherall P, Marks KM, Jakobsson M, Schmitt T, Tani S, Arndt JE, Rovere M, Chayes D, Ferrini V, Wigley R (2015) A new digital bathymetric model of the world's oceans. Earth Space Sci 2:331–345

Remote Sensing for Natural or Man-Made Disasters and Environmental Changes

Alessandro Novellino, Colm Jordan, Gisela Ager, Luke Bateson, Claire Fleming and Pierluigi Confuorto

Abstract Natural and man-made disasters have become an issue of growing concern throughout the world. The frequency and magnitude of disasters threatening large populations living in diverse environments, is rapidly increasing in recent years across the world due to demographic growth, inducing to urban sprawls into hazardous areas. These disasters also have far-reaching implications on sustainable development through social, economic and environmental impact. This chapter summarises three scientific contributions from relevant experiences of the British Geological Survey and the Federico II University of Naples, where remote sensing sensors have been playing a crucial role to potentially support disaster management studies in areas affected by natural hazards. The three cases are: the landslide inventory map of St. Lucia island, tsunami-induced damage along the Sendai coast (Japan) and the landslide geotechnical characterization in Papanice (Italy). For each case study we report the main issue, datasets available and results achieved. Finally, we analyse how recent developments and improved satellite and sensor technologies can support in overcoming the current limitations of using remotely sensed data in disaster management so to fully utilize the capabilities of remote sensing in disaster management and strength cooperation and collaboration between relevant stakeholders including end users.

Keywords Earth observation · Geohazards · Landslide · Tsunami

1 Introduction

In 2016, 342 disasters triggered by natural hazards were registered with 564.4 million people reported affected and US$154 billion estimated cost of damages [4].

A. Novellino (✉) · C. Jordan · G. Ager · L. Bateson · C. Fleming
British Geological Survey, Nicker Hill, Keyworth NG12 5GG, UK
e-mail: alessn@bgs.ac.uk

P. Confuorto
Federico II University of Naples, Largo San Marcellino 10, 80138 Naples, Italy

T. S. Durrani et al. (eds.), *Geological Disaster Monitoring Based on Sensor Networks*, Springer Natural Hazards, https://doi.org/10.1007/978-981-13-0992-2_3

Spawned by the need to rapidly collect vital information to support damage assessment and to assist evaluation and rehabilitation plans, technology innovations have often helped academics, planners, practitioners, policy makers and local communities to assess the potential impact of disasters more efficiently and rapidly, and to track and monitor progress of prevention operations. One technology, which has had an enormous impact on the whole disaster management cycle (response, recovery, mitigation and preparation stage), has been remote sensing. An increasing number of studies have elaborated on the importance and applications of remote sensing in disaster management [1]. A major reason for the adoption of remote sensing is that it is one of the fastest means of acquiring data in timely and cost effective manner upto regional-scale during pre-disaster and post-disaster studies [13], with recent international collaborations and programs exploiting remote sensing technologies to rapidly assist hazard science and response [7]. For example, ARIA from JPL and Caltech, Copernicus from the European Space Agency (ESA) and Sentinel-ASIA from the Asia-Pacific Regional Space Agency Forum.

As a matter of fact, the field of satellite remote sensing has registered massive developments as a result of sensible technological improvements, which led to higher spatial resolution optical and radar systems, hyperspectral sensors, important by-products such as Interferometric Synthetic Aperture Radar (InSAR) ground deformation maps [9] and development of new processing techniques using machine learning systems, able to access and deal with large volumes of data [8].

This chapter contains a collection of three case studies focussed on different application of Earth Observation (EO) techniques for investigating geohazards: St. Lucia and Sendai case studies conducted by the British Geological Survey (BGS) and the Papanice case study from the University of Naples (Italy).

2 St. Lucia Case Study

Due to its location in the Atlantic hurricane belt, the St. Lucia island is prone to hurricanes. Its volcanic terrains [14], together with heavy rainfalls, are inclined to floods and landslides. Also, being placed at the border of a tectonic plate, it faces the threat of earthquakes and tsunamis. Moreover, sea level rise is expected to affect Caribbean countries' coastlines. Consequently, St. Lucia, along with the other small island Caribbean nations, has to carefully determine areas for further development to avoid the creation of new risks. In the framework of the European Space Agency (ESA) *eoworld2* initiative, EO-derived thematic layers have been delivered to the World Bank by BGS using a combination of satellite EO data, ground validation exercises and pre-existing data [3].

The land use map at 2 m resolution was predominantly mapped through a combination of automated classification and visual interpretation of high resolution Pleiades satellite imagery (acquired between 2013 and 2014) and integration, over cloud covered areas, of RapidEye satellite imagery (acquired 2010–2014) and existing land use data (Fig. 1).

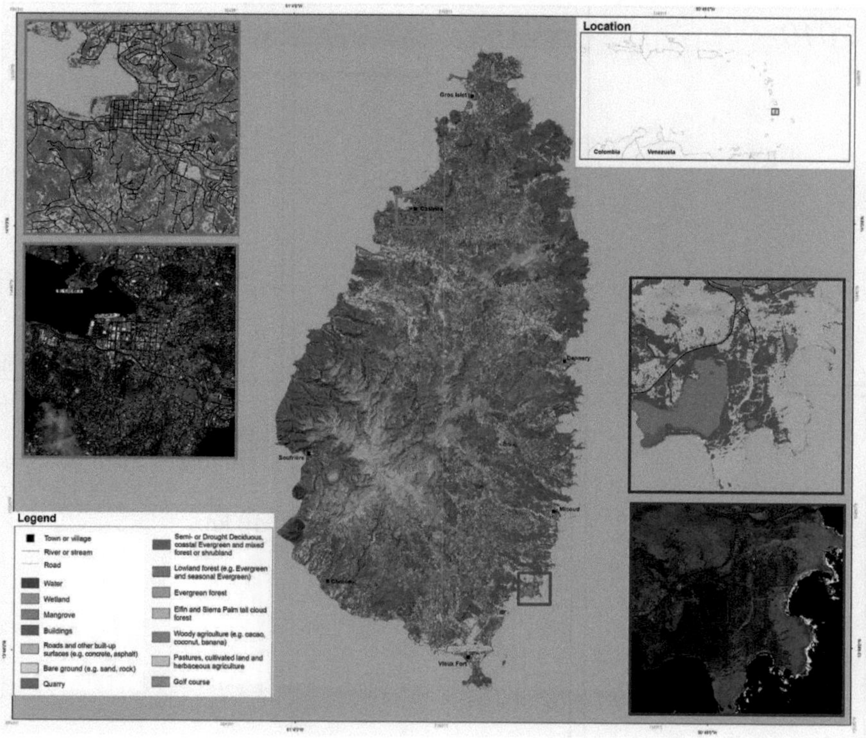

Fig. 1 The land use maps generated for Saint Lucia (modified from [6])

In Saint Lucia the landslide inventories have been typically produced using conventional field surveys. However, if landslides are taking place in remote areas, inventories often cannot provide a reliable and full catalogue of the spatial distribution of event occurrences. Thus, relatively few of them are being recorded and monitored over time. The EO-based multi-temporal landslides inventory created for this project covered the island of Saint Lucia for the years of 2011–2014 (Fig. 2). As a result, 1233 landslide polygons have been detected (compared to the 712 events recorded in the 1995 inventory) and the resulting time series captured two major trigger events: the 2010 Hurricane Tomas and the 2013 Christmas Trough. The study has found that many of the smaller rural roads in the interior were dramatically affected by both events. Moreover, landslides triggered by Hurricane Tomas were rapidly covered by vegetation indicating a quick rate of recovery of the landscape, although, many events were re-activated during the Christmas Trough indicating an increased sensitivity of the landscape to disturbance, and thus higher disaster risk. Extending the multi-temporal record with new acquisitions allowed to create further insights into landscape response to the triggers and this will be vital in establishing relevant hazard and risk assessments.

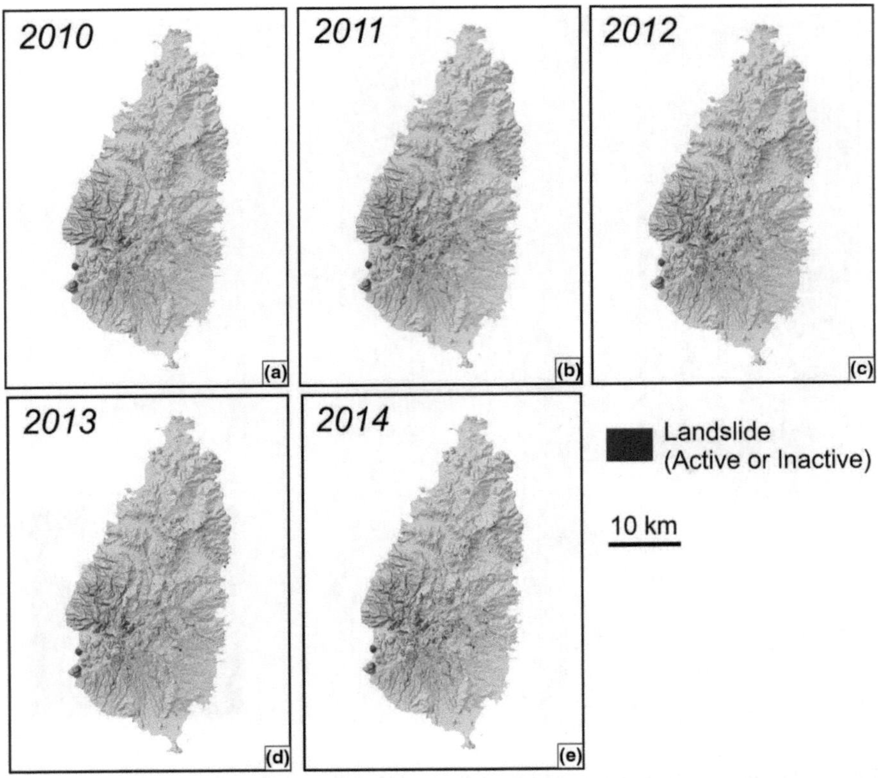

Fig. 2 Landslide inventory maps of St. Lucia showing the distribution of active landslides (modified from [6])

The project supported capacity development within government departments, among land owners and the general public. Several workshops on the use of geo-data for landslide and flood hazard and risk assessment were organized and included a range of stakeholders from World Bank to international development agencies.

3 Papanice Case Study

Landslides in Italy are among the primary cause of death caused by natural hazards [5], as well as of economic losses. Monitoring of such phenomena is of capital importance for public administrations, aiming at the reduction of potential risks. In Papanice, a small hamlet in the province of Crotone (Southern Italy), the post-failure phase of a complex landslide, in the aftermath of a severe rainfall (185 mm in the three days preceding the activation, on February 23, 2012) has been remotely monitored by means of multi-temporal InSAR, which provided ground displacement estimates at millimetre precision from TerraSAR-X images, acquired between October 2013 and October 2014 [2]. Such analysis has been integrated with geomorphological surveys and in situ measurements (boreholes, inclinometers and piezometers), leading to a complete characterization of the landslide. InSAR processing showed a deformation pattern along the main scarp of the landslide, with average displacement of −40 mm/year (along the radar Line of Sight—LOS) in the NW sector, of about −35 mm/year in the central sector and mean values of −5.8 mm/year in the SE sector (Fig. 3).

Time series analysis confirms that the stability of the slope is regulated by piezometric level, depending, in turn, by rainfall data: a slowdown of the movement is reached when the groundwater is at 0.8 m below the ground surface, thus confirming the connection between groundwater presence and triggering/reactivation of

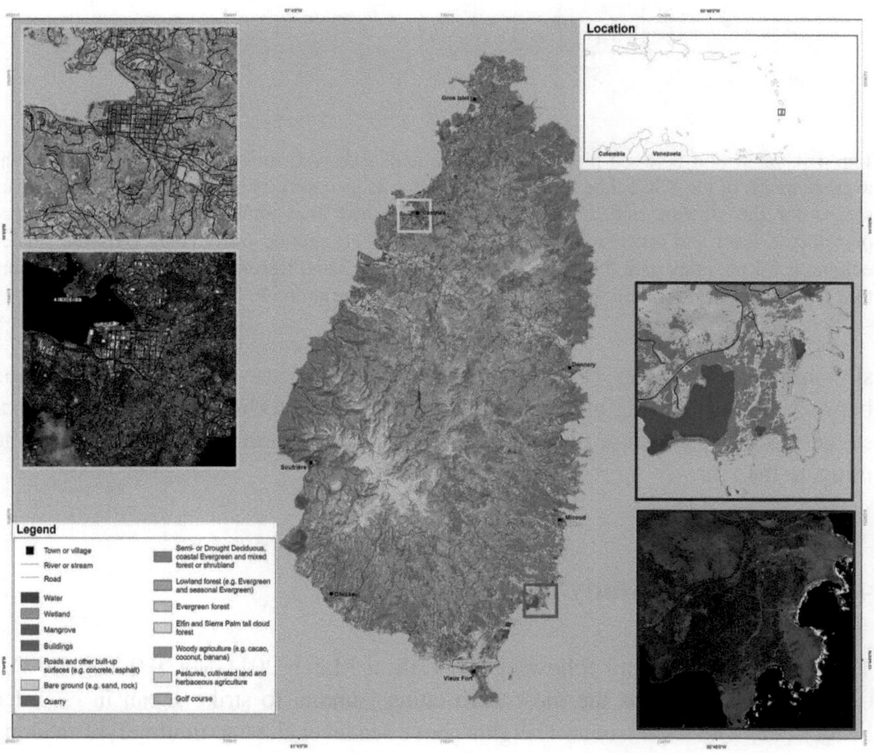

Fig. 3 Displacement rate map obtained with the InSAR processing for the time interval 2013–2014. In the purple rectangle, a zoom in of the area of interest (b). Landslides have been reported in brown-dashed line. The point TSC2 has been selected for the time series analysis. The piezometer location is indicated by the purple triangle, the inclinometer by the pink square (modified from [2] (Color figure online)

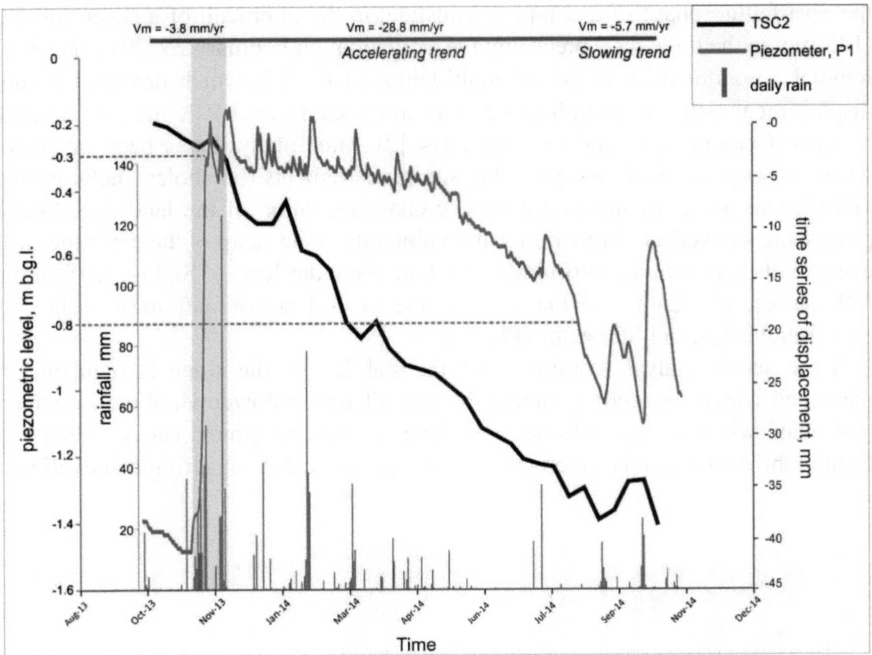

Fig. 4 Comparison between piezometric and rainfall data with time series. On the top bar, the average trends of the TSC2 time series (see Fig. 3 for its location). The average LOS velocities are shown for the slowing and accelerating trends. Dashed lines show the connection between piezometric level and acceleration and deceleration of the ground movement. The blue shaded column is for the rainy period between November 11, 2013 and December 4, 2013, correspondent to a rapid rise of piezometric level (from [2]) (Color figure online)

slow-moving landslides (Fig. 4). For the Papanice case study, the integration between remote sensing and traditional geological/geotechnical investigations has suggested the installation of a superficial drainage system as best intervention to mitigate the risk.

4 Sendai Case Study

The Tohoku-oki tsunami following the 11th of March 2011 Mw 9.0 Tohoku-Oki earthquake (Japan) was the most devastating tsunami to strike Japan in recorded history [11] with local tsunami heights above sea level at a maximum of 40 m along the coast of northern Honshu, where offshore seabed gradients are steep, and funnelling along coastal valleys resulted in intense focussing of the wave [10].

For this case study, we used multi-temporal high-resolution satellite data to map changes in coastal morphology and sedimentary regime from tsunami impact along a 15 km section of the Sendai Plain between the Natori and Abukuma rivers

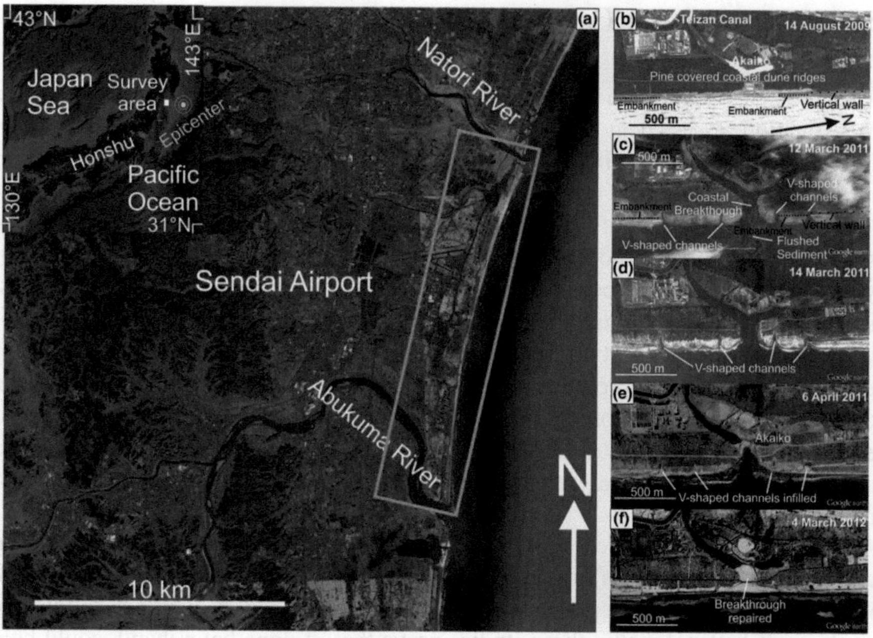

Fig. 5 The Tohoku coast on the island of Honshu, Japan (inset) and study area (red box) along the Sendai coastal plain between the Natori and Abukuma rivers (**a**). Southern sector: time series satellite images of the Akaiko showing the evolution from before the tsunami to the present. **b** 14 August 2009 (pre-tsunami). **c** 12 March 2011. **d** 14 March 2011. **e** 6 April 2011. **f** 4 March 2012 (modified from [12])

(Fig. 5). A specific objective was to identify any differences between those coasts protected by hard engineered structures and those without [12].

In the central part, close to the Sendai Airport, the beach is between 65 and 30 m wide and narrows northward with a continuous protection of tetrapods and concrete blocks along the shoreface along with engineered embankments above the land on either side (Fig. 5a). Since the 2009, satellite images show minor modification of the beach (Fig. 5b). In post-tsunami satellite imagery, the sea has broken through the coastal barrier in the centre of the area and the Akaiko is now connected to the sea through V-shaped channels on the beach with a seaward opening upto 100 m (Fig. 5c). Comparison of the before and after tsunami imagery indicates that there is little change in the sea level at the beach, the landward limit of the tide line remains the same apart from the coastal breakthrough between Fig. 5b, c, and then the gradual healing of the breach through Fig. 5d, e, to the full repair in Fig. 5f.

The research assessed the tsunami impact on the coast, but also succeeded in mapping the spatio-temporal trajectories of damage and reconstruction activities in the area [12]. Furthermore, the results can provide additional evidences for quantifying the effect of the coastal erosion and protecting low-lying areas from flooding.

5 Discussion and Conclusions

The three case studies show how EO has reached a high level of maturity for the investigation, assessment and monitoring of geohazards at multiple scales and in different geological settings. However, they also prove that field verification remains an essential tool to ascertain the validity of image interpretations.

The increasing trend of EO-based studies is mainly driven by the cascade of new spaceborne/airborne sensors and techniques developed especially during the last ten years, which has been possible with the advent of better sensing technologies, improvements in computational resources for running large-scale model simulations and easy accessibility to knowledge and data fostered by the internet with, for example, open-source programs. These processes have consequently enabled the collection, storage, and processing of EO data on crowd-sourced and distributed environments such as cloud platforms, where they can be more effectively analysed and easily shared on a real-time basis through web technologies, social media, and mobile devices.

All these recent advances, have improved the capabilities of EO data acquisition and analysis, and are resulting in the development of 'best-practice' environmental information accessible by planners and decision makers, which enable us to manage geohazards better. Finally, these studies prove that, if done beforehand, could help in identifying particularly vulnerable areas in other parts of the world that could be prone to similar failures and enabling efficient monitoring solutions.

Acknowledgements This publication benefited from inputs and contributions from the following: D. Calcaterra, D. Di Martire, M. Ramondini (University of Naples) and Hussain E. (British Geological Survey). Thanks also to the editors of the special issue in which this chapter is published, for their guidance and patience. Data for the St. Lucia case study have been obtained through the ESA *eoworld2* initiative.

References

1. Bello OM, Aina YA (2014) Satellite remote sensing as a tool in disaster management and sustainable development: towards a synergistic approach. Procedia Soc Behav Sci 120: 365–373
2. Confuorto P, Di Martire D, Centolanza G, Iglesias R, Mallorqui JJ, Novellino A, Plank S, Ramondini M, Thuro K, Calcaterra D (2017) Post-failure evolution analysis of a rainfall-triggered landslide by multi-temporal interferometry SAR approaches integrated with geotechnical analysis. Remote Sens Environ 188:51–72. https://doi.org/10.1016/j.rse.2016.11.002
3. ESA (2016) Earth observation for sustainable development. Partnership Report, September 2016
4. Guha-Sapir D, Hoyois P, Wallemacq P, Below R (2016) Annual disaster statistical review 2016: the numbers and trends. CRED, Brussels

5. Guzzetti F, Mondini AC, Cardinali M, Fiorucci F, Santangelo M, Chang K-T (2012) Landslide inventory maps: new tools for an old problem. Earth Sci Rev 112(1–2):42–66. https://doi.org/10.1016/j.earscirev.2012.02.001
6. Jordan C, Grebby S, Dijkstra T, Dashwood C, Cigna F (2015) Risk information services for disaster risk management (drm) in the Caribbean: operational documentation. British Geological Survey, Nottingham, UK, 69pp. (OR/15/001) (Unpublished)
7. Kaku K, Held A (2013) Sentinel Asia: a space-based disaster management support system in the Asia-Pacific region. Int J Disaster Risk Reduct 6:1–17
8. Lary DJ, Alavi AH, Gandomi AH, Walker AL (2016) Machine learning in geosciences and remote sensing. Geosci Front 7(1):3–10
9. Novellino A, Cigna F, Brahmi M, Sowter A, Bateson L, Marsh S (2017) Assessing the feasibility of a national InSAR ground deformation map of Great Britain with Sentinel-1. Geosciences 7(2):19
10. Shimozono T, Sato S, Okayasu A, Tajima Y, Fritz HM, Liu H, Takagawa T (2012) Propagation and inundation characteristics of the 2011 Tohoku Tsunami on the Central Sanriku Coast. Coastal Eng J 54(1):1250004
11. Simons M, Minson SE, Sladen A, Ortega F, Jiang J, Owen SE, Meng L, Ampuero JP, Wei S, Chu R, Helmberger DV (2011) The 2011 magnitude 9.0 Tohoku-Oki earthquake: Mosaicking the megathrust from seconds to centuries. Science 332(6036):1421–1425
12. Tappin DR, Evans HM, Jordan CJ, Richmond B, Sugawara D, Goto K (2012) Coastal changes in the Sendai area from the impact of the 2011 Tōhoku-oki tsunami: interpretations of time series satellite images, helicopter-borne video footage and field observations. Sediment Geol 282:151–174
13. Williams JG, Rosser NJ, Kincey ME, Benjamin J, Oven KJ, Densmore AL, Milledge DG, Robinson TR, Jordan CA, Dijkstra TA (2018) Satellite-based emergency mapping using optical imagery: experience and reflections from the 2015 Nepal earthquakes. Nat Hazards Earth Syst Sci 18:185–205. https://doi.org/10.5194/nhess-18-185-2018
14. Wohletz K, Heiken G, Ander M, Goff F, Vuataz FD, Wadge G (1986) The Qualibou Caldera, St. Lucia, West Indies. J Volcanol Geotherm Res 27(1–2):77–115

Classification of Post-earthquake High Resolution Image Using Adaptive Dynamic Region Merging and Gravitational Self-Organizing Maps

Aizhu Zhang, Yanling Hao, Genyun Sun, Jinchang Ren, Huimin Zhao, Sophia Zhao and Tariq S. Durrani

Abstract Post-earthquake high resolution image classification has opened up the possibility for rapid damage mapping, which is crucial for damage assessments and emergency rescue. However, the classification accuracy is challenged by the diversity of disaster types as well as the lack of uniform statistical characteristics in post-earthquake high resolution images. In this paper, combining adaptive dynamic region merging (ADRM) and gravitational self-organizing map (gSOM), we propose a novel object-based classification framework. This approach consists of two parts: the segmentation by ADRM and the classification by gSOM. The ADRM produces the homogeneous regions by integrating an adaptive spectral-texture descriptor with a dynamic merging strategy. The gSOM regards the regions as basic unit and characterized them explicitly by fractal texture to adapt to various disaster types. Subsequently, these regions are represented by neurons in a self-organizing map and clustered by adjacency gravitation. By moving the neurons around the gravitational space and merging them according to the gravitation, the gSOM is able to find arbitrary shape and determine the class number automatically. To confirm the validity of the presented approach, three aerial seismic images in Wenchuan covering several disaster types are utilized. The obtained quantitative and qualitative experimental results demonstrated the feasibility and accuracy of the proposed seismic image classification method.

A. Zhang · Y. Hao · G. Sun
School of Geosciences, China University of Petroleum, Qingdao, China
e-mail: genyunsun@163.com

J. Ren (✉) · S. Zhao · T. S. Durrani
Department of Electronic and Electrical Engineering,
University of Strathclyde, Glasgow, UK
e-mail: jinchang.ren@strath.ac.uk

H. Zhao
School of Computers, Guangdong Polytechnic Normal University,
Guangzhou, China
e-mail: zhaohuimin@gpnu.edu.cn

H. Zhao
Guangzhou Key Lab Digital Contents and Security, Guangzhou, China

T. S. Durrani et al. (eds.), *Geological Disaster Monitoring Based on Sensor Networks*, Springer Natural Hazards, https://doi.org/10.1007/978-981-13-0992-2_4

Keywords Earthquake · High resolution image classification · Gravitational self-organizing map · Adaptive dynamic region merging

1 Introduction

Rapid earthquake damage mapping provides rapid, accurate and comprehensive knowledge about the conditions of damaged area, which is vital in the disaster assessment and mitigation [35, 49]. For decades, high resolution remote sensing techniques have been playing an essential role in investigating damage information caused by earthquakes due to its prompt availability after disaster and wide coverage [14]. However, high resolution remote sensing imagery is often characterized by complex data properties in the form of heterogeneity and class imbalance, as well as overlapping class-conditional distributions [7]. Together, these aspects constitute severe challenges for creating damage maps or detecting and localizing damage objects, producing a high degree of uncertainty in obtained results, even for the best performing models.

Object-based image analysis (OBIA) has been firstly used to detect earthquake damage from high resolution remote sensing images in 1998 [19]. Since then, OBIA has been a continual focus in earthquake detection damage using unmanned aerial systems (UAS), LiDAR, and other source data [11, 46, 49]. Sabuncu et al. [41] used image segmentation and condition-based classification to detect damaged buildings following the Ercis-Van earthquake. Similarly, Vetrivel et al. [47] developed a framework for automated mapping of building damage from satellite imagery, using an online classifier that dynamically learns from streaming training samples based on local damage assessments from different sources. In these studies, the feature extraction and classification process are all carried out at an object level, where the objects are obtained by over-segmentation of the satellite image.

Image segmentation is a key and prerequisite step for object-based analysis of high resolution imagery [34]. The resultant segmentation directly influences the final classification accuracy. Region-based methods, such as region merging, have become an increasingly popular image segmentation method [55]. Region merging is able to produce boundary-closed and spatial-continuous regions [56]. Such ability renders it robust for broken patches and speckles. However, in spite of the promising progress achieved in the application of high-resolution imaging, the region merging methods are far from being well-studied in complicated post-earthquake high resolution images, especially for damage objects, such as landslides, debris flow, collapsed buildings and dammed lake. This is mainly because of two problems. The first one is that without the prior knowledge, the similarity of two regions are hard to measure. Another is the static merging strategy always triggers over-segmentation [18] (where some regions can be merged) and under-segmentation [31] (where some regions can be split) problems.

Machine learning can actively adapt and learn the feature representation for objects, often mimicking natural or biological systems, instead of relying on statistical assumptions about data distribution. Learning vector quantization (LVQ) [6], support vector machine (SVM) [30], random forest (RF) [20], and artificial neural networks (ANN) [11] have been attempted to tackle the problem of the post-earthquake VHR image classification. Especially, the gravitational self-organizing map (gSOM) [25] can realize autonomous inference of a rich data representation and able to find arbitrary shape [22, 45, 54]. Therefore, gSOM can be very efficient in handling large and high dimensional datasets and thus shows its potential in the classification of complex post-earthquake high resolution images.

In this paper, we proposed an object-based classification framework that combines adaptive dynamic region merging (ADRM) and gravitational self-organizing map (gSOM). The framework consists of two steps: the segmentation process and the classification process. In the segmentation process, we developed region merging by integrating an adaptive spectral-texture descriptor with a dynamic merging strategy to improve the segmentation efficiency. In this way, ADRM can identify the spatially scattered objects and alleviates the phenomenon of over-segmentation and under-segmentation. The classification process is carried out at an object level, where the objects are obtained by the segmentations. Based on the segmentation results, gSOM produces multiple classifications in view of scale variance in geo-objects. Taking advantage of the diversity of gSOM results, consensus function is then conducted to discover the most suitable classification result.

2 Methodology

In this work, a combined adaptive dynamic region merging (ADRM) and gravitational self-organizing map (gSOM) classification method for post-earthquake high resolution images is proposed. As shown in Fig. 1, the flow diagram of the proposed method includes four major components: (1) initial segmentation: to obtain the over-segmentation results; (2) feature extraction: including histogram-based spectral and spatial feature extraction and adaptive region descriptors, and (3) dynamic region merging, (4) gSOM clustering and clustering ensemble. The first component can be carried out using some well-known segmentation algorithms, such as mean shift [9, 56], level set [43] and super-pixel [56]. In this paper, initial segmentation is obtained by using the mean shift method [9]. The following components are the key parts of the proposed method and will be introduced in details in the next two subsections.

2.1 Feature Extraction

As introduced before, feature of each segmented regions need to be extracted after the initial segmentation. In this paper, the spectral features of regions are descripted by the RGB space quantization while the spatial features are depicted by the local

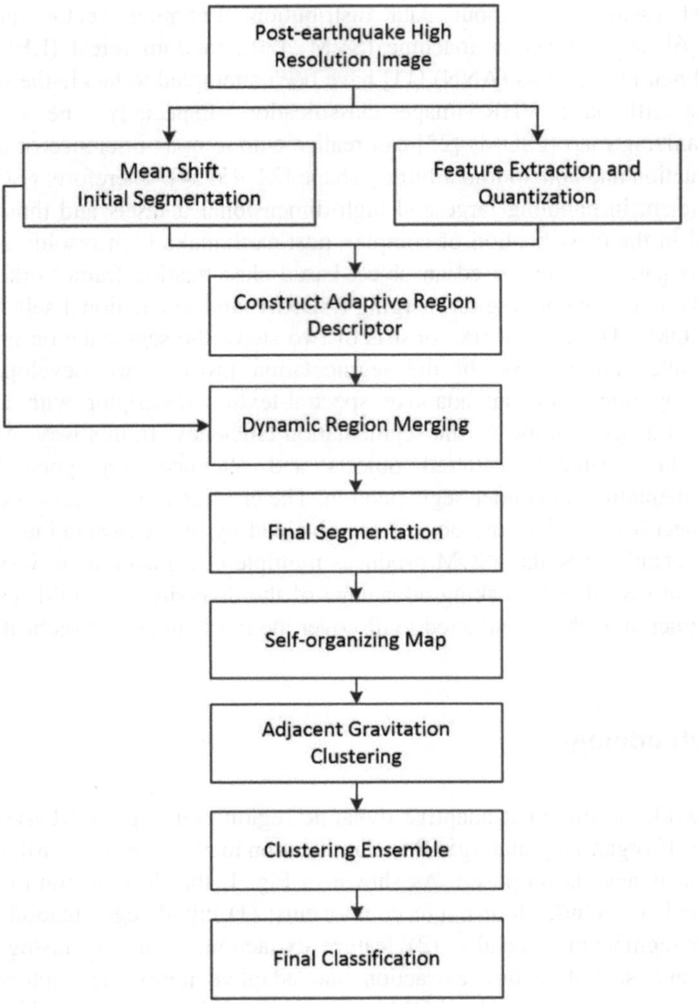

Fig. 1 The proposed classification framework of post-earthquake high resolution image

Gabor texture quantization [5] and global spatial color distribution [28]. Then the spectral and global/local spatial features are combined to construct an adaptive spectral-spatial descriptor to quantify the similarity between regions.

(1) RGB space quantization (RGB) [56]. The RGB space quantization is a kind of well-known robust technique in characterizing the image spectral information. This method can be used to reduce the number of distinct colors and keeps the visual information of the remote sensing image.

(2) **Gabor texture quantization (GAB)** [5]. 2D Gabor texture has shown desirable ability in characterizing the local characteristics of high-resolution remote sensing imagery by adjusting the scales and orientations of the Gabor filter. In this paper, the Gabor texture is adopted to delineate the spatial features of the post-earthquake image. [5, 51]. It is noted that to improve the computation efficiency of subsequent process, not all of the filtered Gabor textures are utilized. Instead only three major components of the Gabor texture feature images are kept by using the principal component analysis (PCA) [1]. Then the three major components are quantified into the Gabor texture quantization feature $\boldsymbol{F}_{GAB}(z)$ using the quantization process similar to RGB space quantization.

(3) **Spatial color distribution (SCD)** [28]. SCD quantifies the spatial distribution of all specific colors by computing the spatial variance of the colors distribution, both horizontally and vertically. In contrast to the Gabor filter, the SCD is a perceptual ancillary can describe the spatial information of remote sensing image in a global perspective and the influence of noise, resolution, and orientation are always very faint [28].

(4) **Fractal texture (FT)**. Texture information is abundant in high resolution post-earthquake images, but very difficult to characterize. Fractal geometry is a powerful texture modeling tool, which is more flexible than Euclidian classical approach to describe and identify a natural and complex object [3]. Thereby, fractal histogram is constructed as follows. The fractal dimension (FD) is firstly estimated for real-world objects through the following expression:

$$D(x) = \lim_{\varepsilon \to 0} \frac{\log(\delta(\varepsilon))}{\log(1/\varepsilon)}, \tag{1}$$

where δ is some kind of complexity measure and ε is the scale over which the measure is taken. Here, we use a particular method called Bouligand–Minkowski [8], in which the texture image is mapped onto a surface, where the i and j coordinates of each pixel $\boldsymbol{I}_{i,j}$ are the horizontal and vertical coordinates of the surface and the intensity value of each pixel in the image are the height Z of the surface. Therefore, this surface is dilated by a sphere with a variable radius ε and, for each radius, the dilation volume $\delta(\varepsilon)$ is computed. In this way, the fractal dimension $\boldsymbol{I}_{FD}$ is commonly estimated by applying Eq. (1). Then sliding window 5×5 for each pixel $\boldsymbol{I}_{i,j}$ is presented to conduct Eq. (1) for the whole image. The local fractal texture features are based on the second-order statistics of fractal dimension, including contrast, dissimilarity, lacunarity, angular second moment, entropy, homogeneity, skewness, inverse difference moment, maximum probability, as shown in Table 1.

Table 1 Fractal texture feature type and meaning

Name	Formula	Meaning
Mean	$\mu_{FD}=\frac{1}{MN}\sum_{i=1}^{M}\sum_{j=1}^{N}I_{FD}(i,j)$	The mean of fractal dimension in a region
Variance	$\sigma_{FD}=\sqrt{\frac{1}{MN}\sum_{i=1}^{M}\sum_{j=1}^{N}\left(I_{FD}(i,j)-\mu_{FD}\right)^2}$	The variance of fractal dimension in a region
Lacunarity	$L_{I_{FD}}=\frac{\frac{1}{MN}\sum_{i=0}^{M}\sum_{j=0}^{N}I_{FD}(i,j)^2}{\left(\frac{1}{MN}\sum_{i=0}^{M}\sum_{j=0}^{N}I_{FD}(i,j)\right)^2}-1$	The lacunarity of fractal dimension in a region.
Angular second moment	$ASM_{FD}=\sum_{i=1}^{M}\sum_{j=1}^{N}\tilde{p}^2(i,j)\log[\tilde{p}(i,j)]$ where $\tilde{p}(i,j)=\frac{1}{\sqrt{2\pi\sigma_{I_{FD}}}}\exp(-\frac{(I_{FD}(i,j)-\mu_{I_{FD}})^2}{2\sigma_{I_{FD}}^2})$	Measure the turbulence of the area, the smaller the value, the more uniform the area, $\tilde{p}(i,j)$ is the normalized probability value of the pixel (i,j) whose coordinates in the fractal dimension image belongs to its own the class.
Contrast	$CON_{FD}=\sum_{n=1}^{L}n^2\left[\sum_{i=1}^{M}\sum_{j=1}^{N}\tilde{p}(i,j)\right],\lvert i-j\rvert=n$	Measure the variance of local area, the smaller the value, the more uniform the texture.
Dissimilarity	$DIS_{FD}=\sum_{i=1}^{M}\sum_{j=1}^{N}\tilde{p}(i,j)\lvert i-j\rvert$	Describes the dissimilarity of pixels and neighboring pixels in the area
Entropy	$ENT_{FD}=-\sum_{i=1}^{M}\sum_{j=1}^{N}\tilde{p}(i,j)\log[\tilde{p}(i,j)]$	Measure the randomness of the pixels in a region, the larger the value, the more complex the texture
Homogeneity	$HOM_{FD}=\sum_{i=1}^{M}\sum_{j=1}^{N}\frac{\tilde{p}(i,j)}{1+\lvert i-j\rvert}\tilde{p}(i,j)$	Describe the uniformity of pixel distribution
Inverse difference moment	$IDM_{FD}=\sum_{i=1}^{M}\sum_{j=1}^{N}\frac{1}{1+(i-j)^2}\tilde{p}(i,j)$	Similar to homogeneity, but more emphasis on the differences between pixels within the region
Maximum probability	$P_{FD}=\max_{ij}(\tilde{p}(i,j))$	Reflect pixel pairs with high probability in a region
Skewness	$S_{FD}=\frac{E(x-\mu_{FD})}{\sigma_{FD}^3}$	Asymmetry of pixel fractal texture values in a region

2.2 *Adaptive Region Descriptor*

After the initial segmentation by mean shift, the homogeneous regions are available. As shown in Fig. 2, the similarity between each two adjacent regions can be

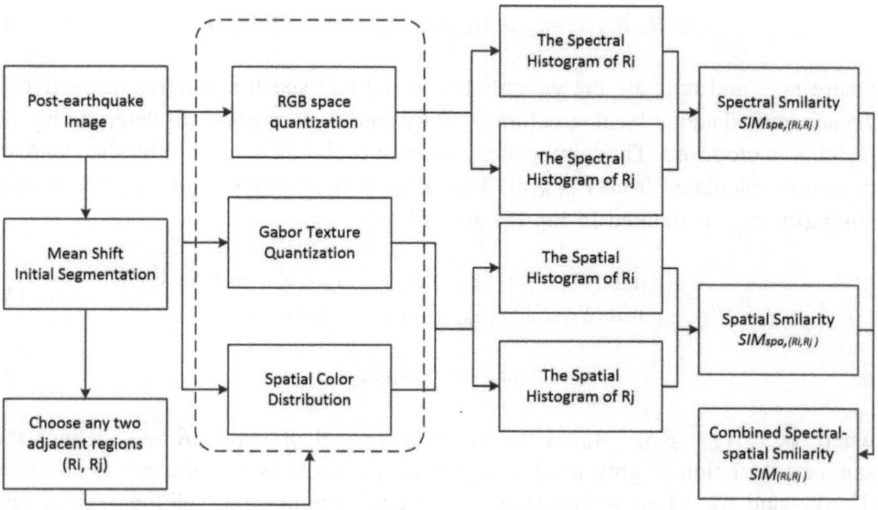

Fig. 2 The feature extraction and the adaptive region descriptor construction

measured by using their corresponding histograms. Assume R_i and R_j are any two adjacent regions in the initial segmentation, and their histograms are extract from both spectral and texture features. It is noted that the spectral histogram $\boldsymbol{H}_{spe,R_i}$ of region R_i is the RGB space quantization histogram $\boldsymbol{H}_{RGB,R_i}$, while the spatial histogram $\boldsymbol{H}_{spa,R_i}$ is obtained by concatenating the Gabor texture quantization histogram $\boldsymbol{H}_{RGB,R_i}$ and spatial color histogram $\boldsymbol{H}_{SCD,R_i}$ as

$$\boldsymbol{H}_{spa,R_i} = \left[\boldsymbol{H}_{GAB,R_i}, \boldsymbol{H}_{SCD,R_i}\right] \tag{2}$$

After obtaining spectral and spatial histograms for the regions, a metric is needed to measure the similarity. Here we use the Bhattacharyya coefficient [10, 27, 36] to measure the spectral and spatial similarities. The spectral similarity between the two adjacent R_i and R_j is defined as follows

$$SIM_{spe}(R_i, R_j) = \sum_{\mu=1}^{Q^3} \sqrt{\boldsymbol{H}^{\mu}_{spe,R_i} \cdot \boldsymbol{H}^{\mu}_{spe,R_j}} \tag{3}$$

According to the Eq. (3), we can find that the larger the Bhattacharyya coefficient of $\boldsymbol{H}_{spe,R_i}$ and H_{spe,R_j} is, the higher their spectral similarity of R_i and R_j is. Similarly, the spatial similarity $SIM_{spe}(R_i, R_j)$ is defined by replacing spectral parts with the spatial counterparts $\boldsymbol{H}_{spa,R_i}$ and H_{spa,R_j}. Then, the adaptive spectral-spatial descriptor $SIM(R_i, R_j)$ is defined by combining the spectral and spatial similarities as

$$SIM(R_i, R_j) = \varpi_{spa} SIM_{spa}(R_i, R_j) + \varpi_{spe} SIM_{spe}(R_i, R_j) \tag{4}$$

where ϖ_{spa} and ϖ_{spe} are the weights for spatial and spectral features respectively. To accommodate the local structures, a fully automatic method for determining the weights is proposed. The homogeneity of each region is evaluated by the standard deviation calculated from $\boldsymbol{F}_{RGB}(z)$. The spectral similarity weight ϖ_{spe} and spatial similarity ϖ_{spa} is defined in Eq. (5) and (6) as

$$\varpi_{spe} = \begin{cases} \max(S_{R_i}, S_{R_j})/(S_{R_i} + S_{R_j}), & S_{R_i} < \psi_i \text{ and } S_{R_j} < \psi_j \\ \min(S_{R_i}, S_{R_j})/(S_{R_i} + S_{R_j}), & otherwise \end{cases} \tag{5}$$

$$\varpi_{spa} = 1 - \varpi_{spe} \tag{6}$$

where $\psi_i = A_{R_i} + \beta_i S_{R_i}$. A_{R_i} is the mean grey level of region R_i, S_{R_i} defines the standard deviation of grey level in region R_i, β_i serves as a coefficient to combine the A_{R_i} and S_{R_i} of R_i to measure the spectral homogeneity of the region, and $\beta_i \in [-2, 2]$ is recommended in Ref. [26].

2.3 Dynamic Region Merging

The ADRM segmentation process addresses region merging in an optimization framework aiming to giving priority to merging the global most similar regions. As illustrated in Fig. 3, ADRM starts from the initially mean shift segmented image, on which a graph-based representation is built. The initially segmented image is first represented by a graph structure. Then the region adjacency graph (RAG) is defined by combining the graph structure with the region similarities. Conventionally, region merging is performed through a full scan of RAG, which is inevitably of high computational cost and low efficiency. To solve the problem, the nearest neighboring graph (NNG) is employed with its superiority in globally finding most similar neighboring regions and dynamically updating the testing order. Therefore, the region merging in ADRM can be conducted in a dynamic style based on the graph models (e.g., RAG and NNG).

Dynamic region merging is conducted according to the testing order of the priority queue in NNG. The priority queue in NNG stores testing order corresponding to the similarity of region pairs in a global view. In the merging process, it is noted that the regions are constantly changed. Consequently, the changed region requires an updated testing order. Traditional region merging adopted the static testing order in the whole procedure, and inevitably causes the inaccurate merging. Instead of the traditional static way, the ADRM adopts a dynamic strategy. Along the changing regions, the graph structure of region partition is updated accordingly. The graph models RAG and NNG are also rebuilt to find globally most suitable solution. Correspondingly, the testing order is dynamically adjusted. In this way,

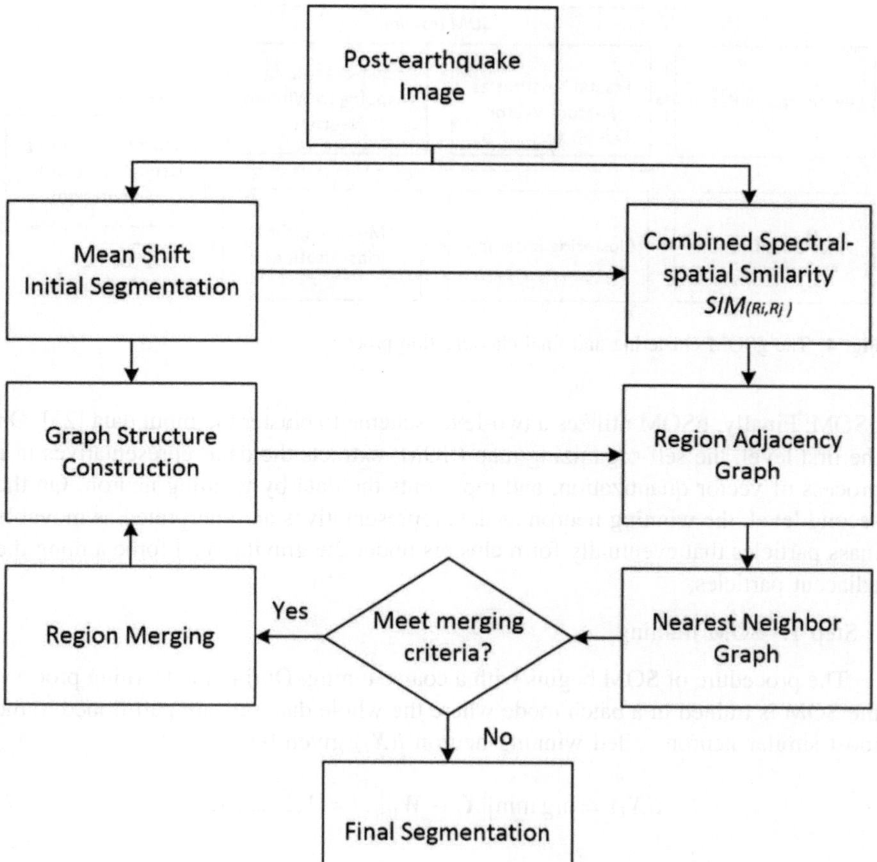

Fig. 3 The adaptive dynamic region merging process

region merging will continue until there is no new merging, that is, there is no cycle or no weight of cycle edge larger than a threshold τ in NNG. Here the parameter τ serves as a scale parameter, which is application-dependent and can be set empirically or interactively.

2.4 gSOM Clustering

In the classification process, the regions in ADRM segmentation are regard as the basic classification units. As illustrated in Fig. 4, at the very beginning, each region is represented by a feature vector. The feature vector is characterized by fractal textures (as shown in Table 1). Then, the feature vectors are transferred to the

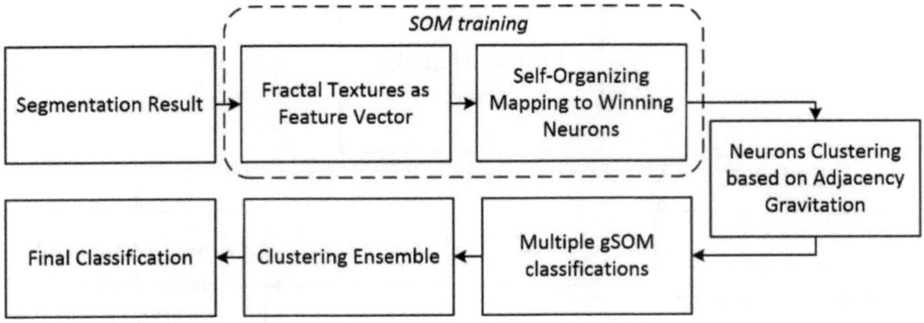

Fig. 4 The gSOM clustering and final classification process

gSOM. Finally, gSOM utilizes a two-level scheme to cluster the input data [25]. On the first level, the self-organizing map (SOM) extracts the data representatives in a process of vector quantization, and represents the data by winning neuron. On the second level, the winning neuron as data representatives are interpreted as movable mass particles that eventually form clusters under the gravitational force among the adjacent particles,

Step 1: SOM training

The procedure of SOM begins with a coarse tuning. During the learning process, the SOM is trained in a batch mode where the whole data sets are partitioned to the most similar neuron called winning neuron $i(\boldsymbol{X}_i)$ given by

$$i(\boldsymbol{X}_i) = \arg\min_j \left\| \boldsymbol{X}_i - \boldsymbol{W}_{ij} \right\|, j = 1, 2, \ldots, l, \tag{7}$$

where $\boldsymbol{X}_i$ is input vector, and $i = 1, 2, \ldots m$ where m is the dimension of input feature vector. $\boldsymbol{W}_{ij}$ is the neuron weight, and $j = 1, 2, \ldots, l$ where l is the neuron number.

Then, the neighbor neuron weights are adjusted according to the winning neuron weight as follows

$$W_i(t+1) = \frac{\sum_{j=1}^{N} h_{i,i(x)}(t) X_j}{\sum_{j=1}^{N} h_{i,i(x)}(t)}, \tag{8}$$

where $h_{i,i(x)}(t)$ is the neighborhood kernel function in t epoch.

Step 2: Neurons clustering based on adjacent gravitation

Based on the trained SOM map, a fine-tuning phase is then conducted to more fully capture the detailed topological pattern of the input data by using a gravitational algorithm and longer training time. During this fine tuning, each codebook vector node converges to an optimal neighbor based on the largest gravitational force. $\vec{x}, \vec{y}$ are two neighboring winning neurons as the data represents are clustering under the adjacent gravitational force

$$\vec{x}(t+1) = \vec{x}(t) - \frac{G(\xi)}{\left\|\vec{d}\right\|^2} \cdot \frac{\vec{d}}{\left\|\vec{d}\right\|}, \tag{9}$$

$$\vec{y}(t+1) = \vec{y}(t) + \frac{G(\xi)}{\left\|\vec{d}\right\|^2} \cdot \frac{\vec{d}}{\left\|\vec{d}\right\|}. \tag{10}$$

where is the distance between $\vec{x}, \vec{y}$. $G(\xi)$ is the gravitational constant variables and ξ decreases in each epoch according to formula $G(\xi+1) = (1 - \Delta G)G(\xi)$ to avoid the scenario that all the particles will collapse into the same point. ΔG is a constant.

2.5 *Clustering Ensemble*

gSOM generates the classification results in an unsupervised way. During this process, there is no explicit teacher or a priori knowledge on the class membership of the samples. The classifier knows the class appearance inexplicitly. Consequently, the autonomy of the classification comes at the expense of more complexity and less reliability. To tackle this problem, clustering ensemble is simple but effective way. Consensus functions [42], including cluster-based similarity partitioning algorithm (CSPA), Meta-clustering algorithm (MCLA) and hypergraph partition algorithm (HGPA), are adopted and chosen according to the principle of the least cost to ensemble the various results, and the final classification is obtained.

3 Experiments

3.1 *Survey Area and Data Description*

The experiments were conducted on three post-earthquake high resolution images, including two subarea images and one urban images. Two test subarea images T1, T2 (the spatial resolution is 0.67 m) were acquired three days after a violent Ms 8.0 earthquake stricken in Wenchuan, China on 12 May 2008 captured by RGB sensors mounted on aerial platforms. The earthquake was centered at approximately 30.98○N and 103.36○E. The focal depth of this earthquake was 14 km and the earthquake devastated a huge area in Wenchuan County. In addition, an urban images T3 (the spatial resolution is 0.8 m) were captured by BJ-2 satellite near the Iran-Iraq border in northwest Iran where an earthquake with a moment magnitude

Table 2 The list of test images

Images	Platform	Size (pixel)	Resolution (m)	Landscape
T1	Aerial	400 × 400	0.67	Rural collapsed residential area, forest, road
T2	Aerial	600 × 600	0.67	Landslides, rural collapsed residential area, forest, farmland
T3	BJ-2	401 × 401	0.80	Urban collapsed building area

of 7.3 occurred on 12 November 2017. The earthquake was centered at approximately 34.90 °N and 45.75 °E, while the focal depth of this earthquake was 20 km. With at least 630 people killed (mostly in Iran and including at least 10 in Iraq) and more than 8100 injured, as well as many more unaccounted for, it is currently the deadliest earthquake of 2017. These test images cover a variety of damage objects such as landslides, debris flow and collapsed residential sites as highlighted in Fig. 2a. Some of the study areas contain certain portions of collapsed residential buildings, while others are partly covered by the landslides or debris flow. That is to say, the mapping of the selected damaged areas is difficult and challenging. The reasons of choosing these test sites mainly are to ensure the variety of secondary disasters types (Table 2).

3.2 *Experiment Setups*

In order to verify the efficacy of the proposed technique, we carried out two groups of experimental validations, including visual assessment and quantitative evaluation. The experiments are implemented in two parts: the ADRM segmentation evaluation and the gSOM classification evaluation.

3.2.1 ADRM Segmentation Evaluation

For the ADRM segmentation evaluation, four powerful and widely used segmentation algorithms, the fractal net evolution approach (FNEA) [2], the statistical region merging (SRM) [40], the classical region growing method (JSEG) [13], and the effective graph-based segmentation (GSEG) [16] were selected to benchmark with the proposed ADRM. FNEA has demonstrated exceptional performance in segmentation of high resolution remote sensing images [23]. SRM exhibits efficient performance in solving significant noise corruption and does not depend on the data distribution [29]. JSEG has a superior robustness in region growing based color-texture image segmentation [13]. GSEG can produce segmentations that obey the global properties of being neither too coarse nor too fine [16]. The parameters are set as detailed in [44]. The ground truth segmentations of three test images were manually interpreted in commercial software eCognition [4] by different experienced experts.

Here, three well-known quantitative performance metrics were used for a detailed evaluation, including Variation of Information (VoI) [33], Global Consistency Error (GCE) [32], Boundary Displacement Error (BDE) [17]. A lower BDE, GCE or VoI value indicates a better result of segmentation.

- Visual Inspection

The visual assessment is to evaluate the segmentation results by visual judgment in both global and detailed view. These three images cover a complicated area that consists of rural and urban collapsed residential area or landslide landscape types. All the damage areas were characterized by severe landscape fragments varied in size and shape, which presented the sharp difference in local spectral and texture. Figures 5, 6, 7, 8 show the segmentation results of T1–T3 by five methods. In each figure, (a)–(e) are the segmentation results by FNEA, JSEG, GSEG, SRM and ADRM respectively, while (f) is the corresponding ground truth segmentation.

According to Figs. 5 and 6, for T1 and T2, GSEG generated under-segmentation for objects of low inter-class variability, such as intact buildings in the areas of the smooth collapsed building ruins, as shown in circle of Fig. 5c and in the little circle of Fig. 6c. In contrast, for T3, GSEG suffered from serious over-segmentation

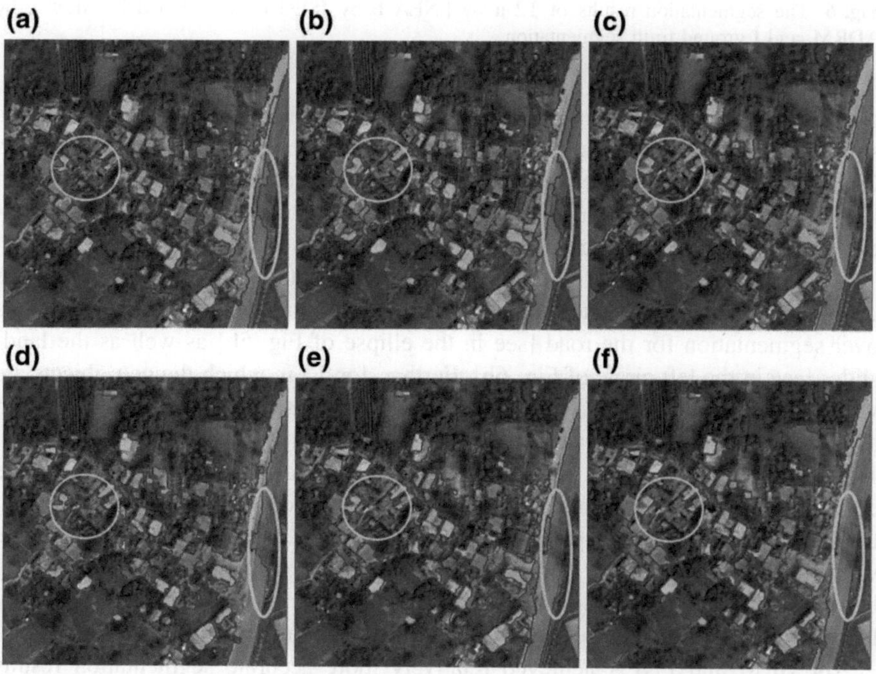

Fig. 5 The segmentation results of T1 **a** by FNEA **b** by JSEG **c** by GSEG **d** by SRM **e** by ADRM, and **f** ground truth segmentation

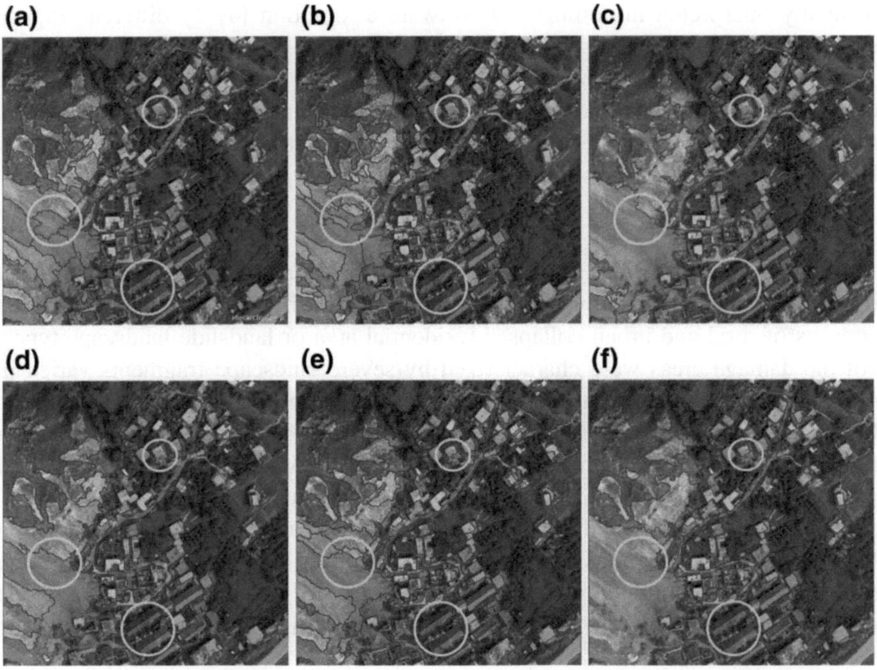

Fig. 6 The segmentation results of T2 **a** by FNEA **b** by JSEG **c** by GSEG **d** by SRM **e** by ADRM, and **f** ground truth segmentation

problem for the objects with high intra-class variability like the intact buildings surrounded by collapsed ruins shown in the middle circle of Fig. 7c. Moreover, GSEG produced serious speckles and noises as illustrated in left and right circles of Fig. 7c. This is because that GSEG only uses the spectral feature without considering the spatial information leading to inadequate description of the objects.

Compared with GSEG, JSEG produced more detailed segmentation for the geo-objects with low inter-class variability, but tended to induce serious over-segmentation for the road [see in the ellipse of Fig. 5b] as well as the landslides [see in the left circle of Fig. 6b]. Further, for T3 in which the geo-objects are varied in components, JSEG triggered under-segmentation for minor objects, such as intact buildings, as illustrated in the little circles in Figs. 6b and 7b. Additionally, JSEG suffered from severe boundary location errors, e.g. parts of the forest were confused with the shadow area as highlighted in left circle of Fig. 7b. This is due to the fact that the relevant spatial information in JSEG generated from the image windows crosses multiple regions, and herein causes difficulty in localizing boundaries of objects. Also, GSEG and SRM have the difficulty of boundary localization when dealing with T1 as exhibited in the ellipse of Fig. 5b.

The SRM and FNEA achieved relatively more accurate segmentation results than GSEG and JSEG as displayed in the Fig. 5a, d. However, they produced some broken patches in the collapsed ruins and the homogeneous road and failed to

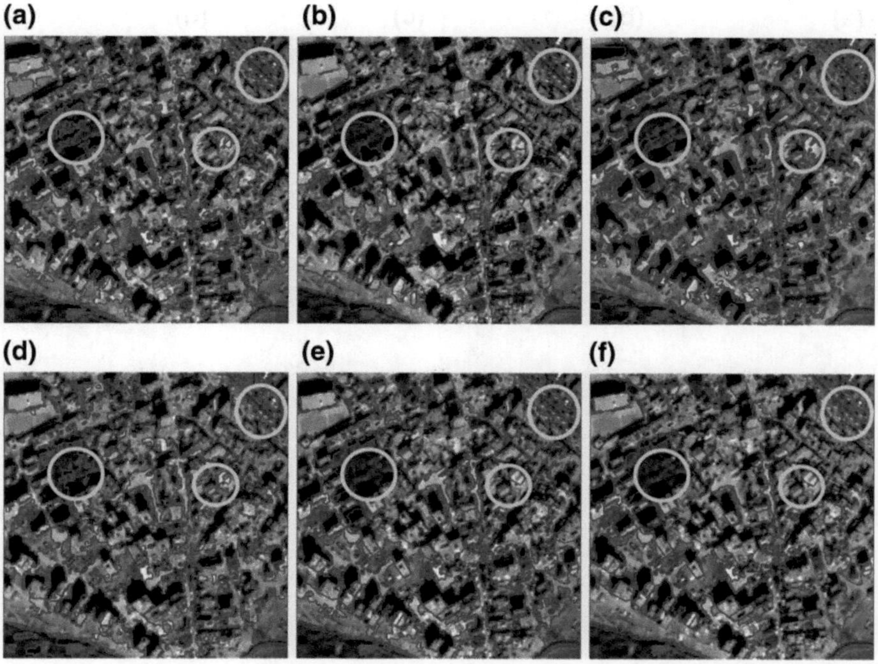

Fig. 7 The segmentation results of T3 **a** by FNEA **b** by JSEG **c** by GSEG **d** by SRM **e** by ADRM, and **f** ground truth segmentation

extract the entity of large geo-objects as shown in the ellipse of Fig. 5a, d, and also in the left circle of Fig. 7a, d. This results from the static testing order in the merging processing adopted by these two methods which is unable to adapt to the changes induced by merging regions.

Owing to the combination of the adaptive spectral–spatial descriptor and the dynamic testing order, ADRM obtained more distinguished results than the other four compared methods in both high intra-class and low inter-class variability areas as presented in Figs. 5, 6, 7e. Furthermore, ADRM is the only algorithm that has completely extracted the areas of collapsed buildings [circle in Fig. 5e] and the road [left circle in Fig. 7e]. To conclude, all these experimental results have demonstrated the superiority and feasibility of the proposed ADRM for post-earthquake high resolution image segmentation.

- Quantitative Evaluation

Table 3 exhibits the quantization evaluations of the segmentation results by five methods. Mean is the average value of metrics. The BDE and GCE errors penalize the under-segmentation problem. As one can see in Table 3 that the average BDE error for ADRM is 8.957, which is a slight improvement over the error of 11.775 achieved by JSEG, 10.401 by FNEA, 11.990 by GSEG and 11.330 by SRM. Yet for T2, the GCE of FNEA is lower than that of ADRM. This is mainly due to the

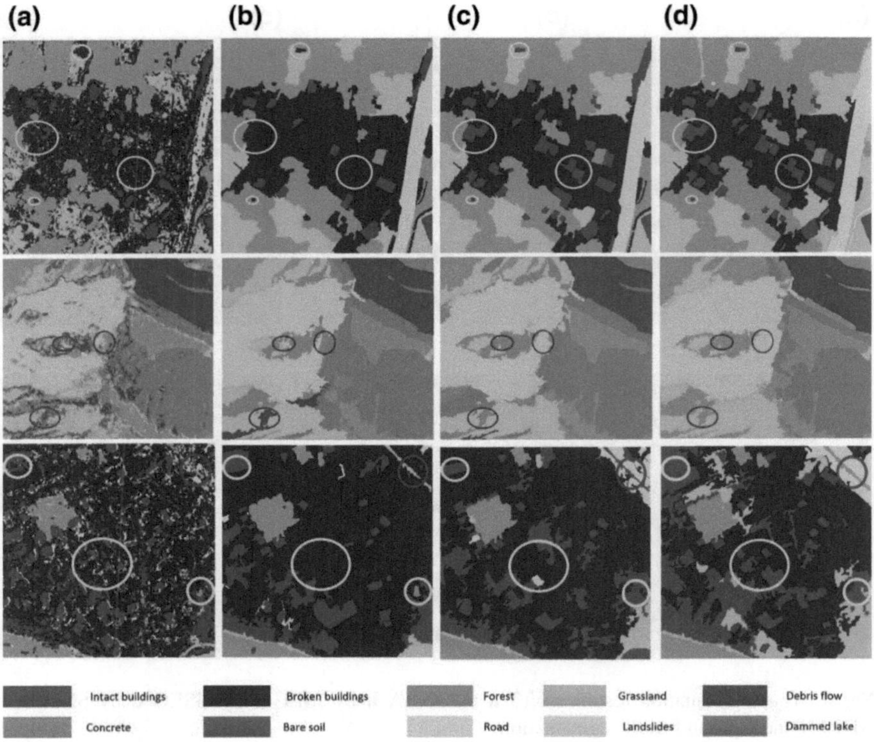

Fig. 8 The classification results of three images **a** by SVM **b** by ADRM_SVM **c** by ADRM_gSOM, and **d** ground truth classifications. From top to down is corresponding to the classification results of T1, T2, T3, respectively

fact that T2 has relatively low-variability image regions, in which FNEA can produce more edge details, resulting lower GCE compared to ADRM. However, the other two metrics of ADRM is better than that of FNEA which indicates its superior segmentation results.

The VoI errors show the degree of the correlation with the extracted ground truth. Thus, it can be considered to be an objective indicator of the image segmentation performance. ADRM achieves the best average value of VoI errors on the six test images, which indicated its better consistency with the ground truth.

In addition, almost all the variances of GCE, VoI and BDE over the entire test images of ADRM are lower than those of the other four methods, which demonstrates the robustness of the ADRM approach to some extent.

3.2.2 gSOM Classification Evaluation

In the classification part, the parameter of the proposed method including the SOM map size *l*, and it was set as 17 in this paper. To be convenient, the proposed

Table 3 Quantitative evaluations of the segmentation results

Metrics	Methods	Images			
		T1	T2	T3	Mean
GCE	JSEG	0.349	0.295	0.545	0.396
	FNEA	0.256	**0.078**	0.231	0.188
	GSEG	0.402	0.237	0.381	0.340
	SRM	0.442	0.198	0.335	0.325
	ADRM	**0.168**	0.228	**0.030**	**0.142**
VoI	JSEG	3.852	2.725	3.905	3.494
	FNEA	3.281	3.784	3.140	3.402
	GSEG	3.209	2.457	4.340	3.335
	SRM	3.188	2.719	3.927	3.278
	ADRM	**2.672**	**2.379**	**1.744**	**2.265**
BDE	JSEG	4.028	27.282	4.015	11.775
	FNEA	2.974	25.809	2.421	10.401
	GSEG	5.579	27.186	3.205	11.990
	SRM	3.332	27.731	2.927	11.330
	ADRM	**2.434**	**22.745**	**1.692**	**8.957**

Bold values indicate the best results in the group

method is denoted by ADRM_gSOM. While the support vector machine (SVM) [12], an advanced supervised kernel classification approach, was selected to compare with the proposed method. The RBF kernel function was selected, and the parameter Gamma in kernel function was set to 0.015. To have a fair comparison, SVM also implemented based on the segmentation by ADRM to generate the object-based classification to illustrate the gSOM efficiency, and denoted by ADRM_SVM.

For the gSOM classification evaluation, confusion matrixes are used to show the performance of final classification, which is shown in Table 4. The ground truth classifications of three test images were further manually interpreted in commercial software eCognition [4] by different experienced experts.

- Visual Inspection

From the Fig. 8, we can find that SVM classification results showed a serious salt-and-pepper appearance throughout the study area, whereas the classification maps of ADRM_SVM and ADRM_gSOM were much more homogeneous and similar to ground truth.

Particularly, visually comparing with the ground truths [Fig. 8d], it was easily found that the classification results of SVM have poor object delineation for intact buildings and the roads, resulting in the scattered regions as shown in circles of Fig. 8a. In addition, the ADRM_SVM classification results were a significant improvement compared to that of SVM. However, ADRM_SVM classification results ignored a lot of the minor objects, especially the intact buildings in the collapsed ruins, as shown in circles in the first row and the third row of Fig. 8. This is due to the fact that there existed inaccurate localization in the SVM classification.

Table 4 The classification accuracy of T1, T2, T3 by SVM, ADRM_SVM and ADRM_gSOM

Image		SVM				ADRM_SVM				ADRM_gSOM			
		PA	UA	OA	K	PA	UA	OA	K	PA	UA	OA	K
T1	Forest	70.34	93.29	65.95	0.53	**94.07**	**94.61**	87.85	0.83	85.57	84.65	**92.56**	**0.90**
	Broken buildings	79.10	50.90			**99.99**	74.86			96.74	**94.93**		
	Road	41.63	67.47			82.54	**100**			**99.99**	87.41		
	Grassland	58.72	63.11			80.81	**100**			**82.5**	96.91		
	Intact buildings	40.54	57.89			40.35	82.02			**80.11**	**100**		
T2	Grassland	65.27	61.01	72.40	0.66	82.64	79.97	87.42	0.84	**95.31**	**80.08**	**92.49**	**0.91**
	Forest	60.48	66.04			**74.60**	**100**			**74.6**	**100**		
	Debris flow	94.31	67.02			**100**	87.01			**100**	**97.90**		
	Dammed lake	74.31	94.52			**96.61**	**100**			**96.61**	**100**		
	Bare soil	36.29	38.93			66.08	70.23			**97.15**	**92.15**		
	Landslides	82.76	92.31			89.68	96.88			**91.58**	**100**		
	Concrete	73.87	72.92			**93.92**	**83.13**			85.68	80.21		
T3	Intact buildings	**45.69**	21.42	47.39	0.22	45.83	36.26	65.47	0.42	47.71	**90.73**	**74.22**	**0.58**
	Forest	66.19	93.53			**91.71**	96.64			89.88	**100**		
	Broken building	62.16	62.34			**92.46**	66.81			91.39	**72.14**		
	Grassland	11.30	35.77			15.62	78.62			**26.75**	**90.5**		
	Bare soil	22.91	33.64			28.18	**67.21**			**58.8**	63.74		
	Road	27.55	12.73			0.44	2.7			**68.52**	**85.78**		

Bold values indicate the best results in the group

Although the SVM had combined with the ADRM segmentation results, it still had trouble in detecting the objects.

In contrast to the results of ADRM_SVM, the classifications of ADRM_gSOM [Fig. 8c] were more consistent with the ground truths than those from ADRM_SVM. It is worth noting that compared with the classifications produced by SVM, the classifications by ADRM_gSOM show more realistic object shapes with merit of the accurate localization and detailed boundary, especially the collapsed buildings areas with varied sizes [in circles of the first row of Fig. 8c]. The efficient combination of the accurate object identification from gSOM classification and the precise object localization from ADRM segmentations renders ADRM_gSOM a perfect tool to extract the complex damaged objects.

Although ADRM_SVM shows an appearance similar to ADRM_gSOM, where most objects were correctly segmented out because it adopted the same scheme of combining ADRM segmentations as ADRM_gSOM. However, there were a certain misclassifications in some local areas. Some isolated thin and elongated objects such as the roads were confused with their surroundings as illustrated in the magenta circle in the third row in Fig. 8b. Besides, it ignored some objects of interests, such as the intact buildings in the big yellow circles of the first and the third rows in Fig. 8b. At the same time, it confused some damage areas such as landslides into the naked land [in the circles in the second row in Fig. 8b]. This stemmed from the misclassifications in the results of SVM as shown in the circle of the second row in Fig. 8a.

- Quantitative Evaluation

Table 4 shows the quantitative evaluation results of T1, T2, T3 by SVM, ADRM_SVM and ADRM_gSOM. OA represents overall accuracy, Kappa is the Kappa coefficient, UA means the user accuracy and PA denotes the product accuracy. According to Table 4, we found that both the OA and Kappa of ADRM_gSOM were higher than the other two methods. This indicates the proposed methods achieved the best performance in this experiment. Compared to SVM, the mean OA of ADRM_gSOM dramatically grew by 25.7%, while the mean Kappa significantly increased by 0.33. In comparison with ADRM_SVM, the mean OA of ADRM_gSOM rose by 8%, while the mean Kappa significantly increased by 0.16. The results further demonstrated the efficiency of gSOM classification.

As seen in Table 4, the PAs of forest, concrete and the UAs of road, grass and intact buildings of ADRM_SVM were a little higher than that of ADRM_gSOM. This is due to the fact that ADRM segmentation has successfully extracted these geo-objects. Therefore, though the SVM had serious noises, the ADRM_SVM still filter out the spurious regions. However, the diversity of gSOM classification results causes the uncertainty to some extent to ensure the robustness for various damages objects.

4 Conclusion

Post-earthquake image segmentation has significant implications for the subsequent images analysis, which is of great importance for rapid damage mapping. The diversity of disaster types and the lack of typical statistical characteristics inherit in post-earthquake images easily leads to confusion of various land covers. To overcome such problems, this paper develops a novel object-based classification scheme by combining ADRM with gSOM. ADRM produced the segmentation with homogeneous regions, while gSOM regard the region as basic unit and classify them into multiple classifications. Then multiple classifications are combined by consensus function to obtain final classification.

The efficiency of the proposed method is evaluated in two parts including the segmentation and classification quantitatively and qualitatively. The segmentation was compared with other four state-of-the-art methods, while the classification was in comparison with SVM and ADRM_SVM. The experiments utilized a variety of post-earthquake high resolution images. Although the features of damage objects varies considerably from case to case, the proposed methods can extract them effectively. The experiment results demonstrated that ADRM_gSOM circumvents the inaccurate classification by integrating of the ADRM and gSOM. Especially, when compared with the results from ADRM_SVM, ADRM_gSOM better preserves the entirety of large objects, meanwhile also prevents small objects from mingling with other objects. It can strike a good balance when partitioning varied-size seismic objects using the adaptive descriptor.

For future work, we will apply effective denoising techniques such as singular spectrum analysis [53] for improving the image quality. State-of-the-art feature extraction approaches will also be explored which include sparse representation [37, 38], saliency detection [48, 50] and curvelet transform [37, 38] as well as deep learning [21, 48]. In addition, advanced image classification approaches such as weakly supervised learning [21] and multi-kernel [15] along with fusion of multispectral/hyperspectral [39] and Synthetic Aperture Radar (SAR) [24, 52] images will also be focused.

Acknowledgments This study was funded by the National Natural Science Foundation of China (41471353,61672008), the Natural Science Foundation of Shandong Province (ZR201709180096, ZR201702100118), the Fundamental Research Funds for the Central Universities (18CX05030A, 18CX02179A), the Postdoctoral Application and Research Projects of Qingdao (BY20170204), and Guangdong Provincial Application-oriented Technical Research and Development Special fund project (2016B010127006, 2017A050501039).

References

1. Abdi H, Williams LJ (2010) Principal component analysis. Wiley Interdiscip Rev Comput Stat 2(4):433–459. https://doi.org/10.1002/wics.101
2. Baatz M, Schäpe A (2000) Multiresolution segmentation: an optimization approach for high quality multi-scale image segmentation. Angewandte Geographische Informationsverarbeitung XII:12–23
3. Backes AR, Casanova D, Bruno OM (2012) Color texture analysis based on fractal descriptors. Pattern Recogn 45(5):1984–1992. https://doi.org/10.1016/j.patcog.2011.11.009
4. Benz UC, Hofmann P, Willhauck G, Lingenfelder I, Heynen M (2004) Multi-resolution, object-oriented fuzzy analysis of remote sensing data for GIS-ready information. ISPRS J Photogramm Remote Sens 58(3–4):239–258. https://doi.org/10.1016/j.isprsjprs.2003.10.002
5. Bhagavathy S, Manjunath BS (2005) Modeling and detection of geospatial objects using texture motifs. Geosci Remote Sens IEEE Trans 44(12):3706–3715. https://doi.org/10.1109/TGRS.2006.881741
6. Blaschke T (2010) Object based image analysis for remote sensing. ISPRS J Photogramm Remote Sens 65(1):2–16. https://doi.org/10.1016/j.isprsjprs.2009.06.004
7. Campsvalls G, Bruzzone L (2009) Kernel methods for remote sensing data analysis. https://doi.org/10.1002/9780470748992.ch9
8. Casanova D, Florindo JB, Falvo M, Bruno OM (2016) Texture analysis using fractal descriptors estimated by the mutual interference of color channels. Inf Sci 346–347:58–72. https://doi.org/10.1016/j.ins.2016.01.0771
9. Comaniciu D, Meer P (2002) Mean shift: a robust approach toward feature space analysis. IEEE Trans Pattern Anal Mach Intell 24(5):603–619. https://doi.org/10.1109/34.1000236
10. Comaniciu D, Ramesh V, Meer P (2003) Kernel-based object tracking. IEEE Trans Pattern Anal Mach Intell 25(5):564–577. https://doi.org/10.1109/TPAMI.2003.1195991
11. Cooner A, Shao Y, Campbell J (2016) Detection of urban damage using remote sensing and machine learning algorithms: Revisiting the 2010 Haiti earthquake. Remote Sens 8(10):2072–4292. https://doi.org/10.1109/34.94698510.3390/rs8100868
12. Cortes C, Vapnik V (1995) Support-vector networks. Mach Learn 20(3):273–297
13. Deng Y, Manjunath B (2001) Unsupervised segmentation of color-texture regions in images and video. Pattern Anal Machine Intell IEEE Trans 23(8):800–810. https://doi.org/10.1109/34.946985
14. Dong L, Shan J (2013) A comprehensive review of earthquake-induced building damage detection with remote sensing techniques. ISPRS J Photogramm Remote Sens 84(7):85–99. https://doi.org/10.1016/j.isprsjprs.2013.06.011
15. Fang L, Li S, Duan W, Ren J, Benediktsson JA (2015) Classification of hyperspectral images by exploiting spectral–spatial information of superpixel via multiple kernels. IEEE Trans Geosci Remote Sens 53(12):6663–6674. https://doi.org/10.1109/TGRS.2015.2445767
16. Felzenszwalb PF, Huttenlocher DP (2004) Efficient graph-based image segmentation. Int J Comput Vision 59(2):167–181. https://doi.org/10.1023/b:visi.0000022288.19776.77
17. Freixenet J, Muñoz X, Raba D, Martí J, Cufí X (2002) Yet another survey on image segmentation: region and boundary information integration. In: Proceedings ECCV 2002, vol 2352, pp 408–422. https://doi.org/10.1007/3-540-47977-5_27
18. Gaetano R, Masi G, Poggi G, Verdoliva L, Scarpa G (2015) Marker-controlled watershed-based segmentation of multiresolution remote sensing images. IEEE Trans Geosci Remote Sens 53(6):2987–3004. https://doi.org/10.1109/TGRS.2014.2367129
19. Gamba P (1998) GIS and image understanding for near-real-time earthquake damage assessment. Photogramm Eng Remote Sens 64(10):987–994. http://worldcat.org/issn/00991112
20. Geiß C, Pelizari PA, Marconcini M, Sengara W, Edwards M, Lakes T, Taubenböck H (2015) Estimation of seismic building structural types using multi-sensor remote sensing and

machine learning techniques. Isprs J Photogramm Remote Sens 104:175–188. https://doi.org/10.1016/j.isprsjprs.2014.07.016

21. Han J, Zhang D, Cheng G, Guo L, Ren J (2015) Object detection in optical remote sensing images based on weakly supervised learning and high-level feature learning. IEEE Trans Geosci Remote Sens 53(6):3325–3337. https://doi.org/10.1109/TGRS.2014.2374218
22. Hao Y, Sun G (2016) Gravitational self-organizing map-based seismic image classification with an adaptive spectral-textural descriptor. SPIE Remote Sensing. doi 10(1117/12):2241272
23. Hay GJ, Blaschke T, Marceau DJ, Bouchard A (2003) A comparison of three image-object methods for the multiscale analysis of landscape structure. ISPRS J Photogramm Remote Sens 57(5–6):327–345. https://doi.org/10.1016/S0924-2716(02)00162-4
24. Hwang B, Renn J, McCormack S, Berry C, Ben Ayed I, Graber HC, Aptoula E (2017) A practical algorithm for the retrieval of floe size distribution of Arctic sea ice from high-resolution satellite synthetic aperture radar imagery. Elem Sci Anth 5:38. https://doi.org/10.1525/elementa.154
25. Ilc N, Dobnikar A (2012) Generation of a clustering ensemble based on a gravitational self-organising map. Neurocomputing 96:47–56. https://doi.org/10.1016/j.neucom.2011.10.043
26. Jiangye Y, Wang DL, Li X (2014) Remote sensing image segmentation by combining spectral and texture features. IEEE Trans Geosci Remote Sens 52(1):16–24. https://doi.org/10.1109/TGRS.2012.2234755
27. Kailath T (1967) The divergence and Bhattacharyya distance measures in signal selection. IEEE Trans commun Technol 15(1):52–60. https://doi.org/10.1109/TCOM.1967.1089532
28. Kiranyaz S, Birinci M, Gabbouj M (2010) Perceptual color descriptor based on spatial distribution: a top-down approach. Image Vis Comput 28(8):1309–1326. https://doi.org/10.1016/j.imavis.2010.01.012
29. Lang F, Yang J, Li D, Zhao L, Shi L (2014) Polarimetric SAR image segmentation using statistical region merging. IEEE Geosci Remote Sens Lett 11(2):509–513. https://doi.org/10.1109/LGRS.2013.2271040
30. Li P, Song B, Xu H (2011) Urban building damage detection from very high resolution imagery by one-class SVM and shadow information. In: IEEE international geoscience and remote sensing symposium, Igarss 2011, Vancouver, Bc, Canada, July, 2011. pp 1409–1412. https://doi.org/10.1109/igarss.2011.6049330
31. Liu J, Li P, Wang X (2015) A new segmentation method for very high resolution imagery using spectral and morphological information. ISPRS J Photogramm Remote Sens 101:145–162. https://doi.org/10.1016/j.isprsjprs.2014.11.009
32. Martin D, Fowlkes C, Tal D, Malik J (2001) A database of human segmented natural images and its application to evaluating segmentation algorithms and measuring ecological statistics. In: Computer vision, 2001. ICCV 2001. Proceedings. Eighth IEEE international conference on, 2001. vol. 412, pp 416–423. https://doi.org/10.1109/iccv.2001.937655
33. Meilă M (2007) Comparing clusterings—an information based distance. J Multivar Anal 98(5):873–895. https://doi.org/10.1016/j.jmva.2006.11.013
34. Ming D, Ci T, Cai H, Li L, Qiao C, Du J (2012) Semivariogram-based spatial bandwidth selection for remote sensing image segmentation with mean-shift algorithm. Geosci Remote Sens Lett IEEE 9(5):813–817. https://doi.org/10.1109/LGRS.2011.2182604
35. Nex F, Rupnik E, Toschi I, Remondino F (2014) Automated processing of high resolution airborne images for earthquake damage assessment. ISPRS Int Arch Photogramm Remote Sens Spatial Inf Sci XL-1(1):315–321. https://doi.org/10.5194/isprsarchives-xl-1-315-2014
36. Pudil P, Somol P, Haindl M (1972) Introduction to statistical pattern recognition. Acad Press. https://doi.org/10.1002/0470854774.ch1
37. Qiao T, Ren J, Wang Z, Zabalza J, Sun M, Zhao H, Li S, Yuen P, Zhao H, Benediktsson JA, Dai Q, Marshall S (2017) Effective denoising and classification of hyperspectral images using curvelet transform and singular spectrum analysis. IEEE Trans Geosci Remote Sens 55(1):119–133. https://doi.org/10.1109/TGRS.2016.2598065

38. Qiao T, Yang Z, Ren J, Yuen P, Zhao H, Sun G, Marshall S, Benediktsson JA (2017) Joint bilateral filtering and spectral similarity-based sparse representation: a generic framework for effective feature extraction and data classification in hyperspectral imaging. Pattern Recogn 77:316–328. https://doi.org/10.1016/j.patcog.2017.10.008
39. Ren J, Zabalza J, Marshall S, Zheng J (2014) Effective feature extraction and data redution in remote sensing using hyperspectral imaging. IEEE Signal Process Mag 31(4):149–154. https://doi.org/10.1109/MSP.2014.2312071
40. Richard N, Frank N (2004) Statistical region merging. IEEE Trans Pattern Anal Mach Intell 26(11):1452–1458. https://doi.org/10.1109/TPAMI.2004.110
41. Sabuncu A, Avci ZDU, Sunar F (2016) Preliminary results of earthquake-induced building damage detection with object-based image classification. ISPRS Int Arch Photogramm Remote Sens Spatial Inf Sci 2016:347–350. https://doi.org/10.5194/isprsarchives-XLI-B7-347-2016
42. Strehl A, Ghosh J (2003) Cluster ensembles—a knowledge reuse framework for combining multiple partitions. J Machine Learn Res 3(3):583–617. https://doi.org/10.1162/153244303321897735
43. Sumengen B (2004) Variational image segmentation and curve evolution on natural images. University of California, Santa Barbara
44. Sun G, Hao Y, Chen X, Ren J, Zhang A, Huang B, Zhang Y, Jia X (2017) Dynamic post-earthquake image segmentation with an adaptive spectral-spatial descriptor. Remote Sens 9(9):899. https://doi.org/10.3390/rs9090899
45. Sun G, Ma P, Ren J, Li X, Zhang A, Jia X (2018) A stability constrained adaptive alpha for gravitational search algorithm. Knowl Based Syst 139:200–213. https://doi.org/10.1016/j.knosys.2017.10.018
46. Tiede D, Lang S, Füreder P, Hölbling D, Hoffmann C, Zeil P (2011) Automated damage indication for rapid geospatial reporting. An operational object-based approach to damage density mapping following the 2010 Haiti earthquake. Photogramm Eng Remote Sens 77 (9):933–942. https://doi.org/10.14358/pers.77.9.933
47. Vetrivel A, Gerke M, Kerle N, Nex F, Vosselman G (2017) Disaster damage detection through synergistic use of deep learning and 3D point cloud features derived from very high resolution oblique aerial images, and multiple-kernel-learning. ISPRS J Photogramm Remote Sens. https://doi.org/10.1016/j.isprsjprs.2017.03.001(InPress)
48. Wang Z, Ren J, Zhang D, Sun M, Jiang J (2018) A deep-learning-based feature hybrid framework for spatiotemporal saliency detection inside videos. Neurocomputing 287:68–83. https://doi.org/10.1016/j.neucom.2018.01.076
49. Witharana C, Civco DL, Meyer TH (2014) Evaluation of data fusion and image segmentation in earth observation based rapid mapping workflows. ISPRS J Photogramm Remote Sens 87:1–18. https://doi.org/10.1016/j.isprsjprs.2013.10.005
50. Yan Y, Ren J, Sun G, Zhao H, Han J, Li X, Marshall S, Zhan J (2018) Unsupervised image saliency detection with Gestalt-laws guided optimization and visual attention based refinement. Pattern Recogn 79:65–78. https://doi.org/10.1016/j.patcog.2018.02.004
51. Yang M, Zhang L, Shiu CK, Zhang D (2012) Monogenic binary coding: an efficient local feature extraction approach to face recognition. IEEE Trans Inf Forensics Secur 7(6):1738–1751. https://doi.org/10.1109/TIFS.2012.2217332
52. Zabalza J, Ren J, Yang M, Zhang Y, Wang J, Marshall S, Han J (2014) Novel Folded-PCA for improved feature extraction and data reduction with hyperspectral imaging and SAR in remote sensing. ISPRS J Photogramm Remote Sens 93:112–122. https://doi.org/10.1016/j.isprsjprs.2014.04.006
53. Zabalza J, Ren J, Zheng J, Han J, Zhao H, Li S, Marshall S (2015) Novel two-dimensional singular spectrum analysis for effective feature extraction and data classification in hyperspectral imaging. IEEE Trans Geosci Remote Sens 53(8):4418–4433. https://doi.org/10.1109/TGRS.2015.2398468

54. Zhang A, Sun G, Ren J, Li X, Wang Z, Jia X (2018) A dunamic neighborhood learning-based gravitational search algorithm. IEEE Trans Cybern 48(1):436–447. https://doi.org/10.1109/TCYB.2016.2641986
55. Zhao W, Du S (2016) Learning multiscale and deep representations for classifying remotely sensed imagery. ISPRS J Photogramm Remote Sens 113:155–165. https://doi.org/10.1016/j.isprsjprs.2016.01.004
56. Zhou C, Wu D, Qin W, Liu C (2015) An efficient two-stage region merging method for interactive image segmentation. Comput Electr Eng 54:220–229. https://doi.org/10.1016/j.compeleceng.2015.09.013

A Survey on the Role of Wireless Sensor Networks and IoT in Disaster Management

Ahsan Adeel, Mandar Gogate, Saadullah Farooq, Cosimo Ieracitano, Kia Dashtipour, Hadi Larijani and Amir Hussain

Abstract Extreme events and disasters resulting from climate change or other ecological factors are difficult to predict and manage. Current limitations of state-of-the-art approaches to disaster prediction and management could be addressed by adopting new unorthodox risk assessment and management strategies. The next generation Internet of Things (IoT), Wireless Sensor Networks (WSNs), 5G wireless communication, and big data analytics technologies are the key enablers for future effective disaster management infrastructures. In this chapter, we commissioned a survey on emerging wireless communication technologies with potential for enhancing disaster prediction, monitoring, and management systems. Challenges, opportunities, and future research trends are highlighted to provide some insight on the potential future work for researchers in this field.

A. Adeel (✉) · M. Gogate · S. Farooq · K. Dashtipour · A. Hussain
Department of Computing Science and Mathematics, Faculty of Natural Sciences, University of Stirling, Stirling, Scotland, UK
e-mail: aad@cs.stir.ac.uk

M. Gogate
e-mail: mandarvinayakgogate@gmail.com

S. Farooq
e-mail: saadullahfarooq93@gmail.com

K. Dashtipour
e-mail: twf.kia@gmail.com

A. Hussain
e-mail: ahu@cs.stir.ac.uk

C. Ieracitano
DICEAM, University Mediterranea of Reggio Calabria, Reggio Calabria, Italy
e-mail: cosimo.ieracitano@unirc.it

H. Larijani
School of Engineering and Built Environment, Glasgow Caledonian University, Glasgow, Scotland, UK
e-mail: hla@gcu.ac.uk

T. S. Durrani et al. (eds.), *Geological Disaster Monitoring Based on Sensor Networks*, Springer Natural Hazards, https://doi.org/10.1007/978-981-13-0992-2_5

Keywords Disaster management · Internet of things · Wireless sensor networks 5G wireless communication · Big data analytics

1 Introduction

Albert Einstein once said: The true sign of intelligence is not knowledge but imagination. Imagining how technology could be exploited to address the challenging real-world problems is necessary, especially when the development of new device or system is considered. Disaster management is one of the challenging real-world problems that sought out emerging technologies. Natural disasters have been visiting every part of the globe and the world has become increasingly vulnerable. Nearly 3 million people worldwide have been killed in past 20 years due to natural disasters, including earthquakes, landslides, floods, cyclones, snow avalanches etc. Intelligent infrastructures to enhance disaster management, community resilience and public safety have become inevitably important, with the aim to save lives, reduce risk and disaster impacts, permitting efficient use of material and social resources, and protect the quality of life and economic stability across entire regions. At the leading edge of disaster management initiatives are the collection, integration, management and analysis of an increasingly complex web of multi-modal data and digital information originating from mobile, fixed, and embedded sources. A national program for investment in intelligent infrastructure can achieve dramatic economies of scale and reduce long-term national debt. For example, United States spends billions of dollars annually to suppress catastrophic wildfires, which consume millions of acres each year. A single major hurricane or tornado can claim hundreds of lives and cause billions of dollars damage. Intelligent infrastructure technologies such as computer models taught through machine learning and calibrated on big data, including ground-based sensors, streaming video from unmanned aerial vehicles, and satellite imagery, could significantly reduce the social and economic costs of such disasters. Priority should be placed on the areas such as sensing and data collection, communication and coordination, big data modelling frameworks (including analytics and tools for disaster prediction and management), and social computing.

There is a great deal of interest in developing disasters management systems capable of saving lives, properties, and minimize the costly economic investments. In order to develop an effective monitoring infrastructure, the information has to be gathered from different sources. In this context, IoT technology is reporting considerable success [1]. In last few years, innovative real-time monitoring and disasters warning systems are based on the emerging IoT paradigm, in which things (i.e. sensors) are globally interconnected. WSNs as part of IoT has been employed widely for monitoring natural disasters in remote and inaccessible areas [2–4]. WSNs use autonomous low-energy sensor nodes capable of measuring and recording surrounding environmental conditions. Each sensor node typically consists of a power supply, a micro-controller, a wireless radio transmission and a set of environmental sensors (i.e. humidity, pressure, temperature). WSNs and IoT together with the recent advances in the Information and Communication Technology (ICT) could develop ever more intelligent and connected infrastructures, where a huge amount of data could be gathered and processed [5]. The combination of these heterogeneous resources (gathered from digital infrastructures) and the latest artificial intelligence technology could be used to develop next-generation of disaster management systems.

The rest of the chapter is organized as follows: Sect. 2 first presents an overview of state-of-the-art WSNs driven disaster monitoring and management systems, including emerging technologies such as 5G, Device to Device communication, Fourth Generation (4G)/LTE, and software defined radio. Section 3 presents an overview of existing IoT standards (LoRa/4G LTE), their limitations, and future research directions. Finally, Sect. 4 concludes this chapter.

2 WSN Driven Disaster Monitoring and Management Systems

2.1 Applications of Sensor Networks in Disasters Management

The application of sensor networks for monitoring natural hazards (such as floods [6], wildfires [7] or sandstorms [8]) has become a special research topic for many researchers and engineers. In this context, a lot of work has been focusing on using WSNs for monitoring landslides (e.g. downfalls of a large mass of ground, rock fragments and debris especially in unstable areas where intense rainfalls, floods or earthquakes occur and might cause loss of lives, damage buildings and influence the economy [9]). Kotta et al. [10] proposed a WSNs system based on accelerometers for vibrations detection triggered by landslides. Experimental results showed that accelerometers values above 1 g (gravity) indicated intense mass sliding and hazardous conditions. Ramesh et al. [11] installed a sensors distributed monitoring system based on 50 geological sensors and 20 wireless sensor nodes to monitor a local zone highly at risk from landslides in India (Idukki, Kerala State). The proposed system was able to provide three level alerts (low, intermediate, high) and its effectiveness was tested during the monsoon season. Terzis et al. [12] proposed a sensor columns based network to detect the slip surface location and the trigger of a landslide. Lee et al. [13] presented a slope movement monitoring WSNs system capable of reducing the power consumption during the standby mode (0.05 mA at 3.6 V). Rossi et al. [14] reported the development of a landslides monitoring system installed in the Apennines (North Italy). Similarly, Giorgettiet al. [15] deployed a network of 15 wireless sensors on a landslide in Torgio vannetto (Italy) observing high level of robustness in term of self-organization, node failures, and energy consumption.

2.2 5G and Device to Device Communication

The upcoming 5G systems are envisioned to have the crucial capabilities such as network flexibility, (re)configuration and resilience and therefore, expected to play a key role in improving disaster situation communications. Furthermore, in 5G, network will provide media independent handover (IEEE 802.21 support) allowing seamless hand-off between various available networks thereby enabling disaster communication without any disruption. 5G networks are not only expected to attain much faster transmission throughput, but also support the emerging use-cases related to the IoT, Machine Type Communications (MTC), broadcast-like services and lifeline communications during natural disasters. 5G will fulfill these demands by adopting new technologies like proximity services, through which devices communicate with each other directly instead of relying on base stations (eNodeB) of network operators [16].

Device-to-Device (D2D) communication has also been used in disaster scenarios (e.g. for public safety and warning messages) to manage the radio spectrum and energy consumption for providing high Quality of Experience (QoE) and better Quality of Service (QoS). In a disaster, the effective use of the radio resources is of extreme importance with the goal of serving a large number of affected people to collect information from different nodes in the disaster zone. In this context, D2D communication will be an effective solution allowing an efficient spectrum allocation without adding any further delay in content uploading for the User Equipments (UEs) [16].

2.3 Software Defined Radio

While LTE provides a solution to address the lack of broadband connectivity in disaster network, Software Defined Radio (SDR) technology provides a solution to address the lack of interoperability in a wireless communication scenario e.g., in military applications. SDR enables a platform to interface and communicate with different communication technologies. SDR technology could be used to support various wireless communications technologies on the same radio platform. It is also essential to define a common waveform to support the wireless backbone network. Though SDR is a promising technology, its potential application in the disaster management requires addressing various issues, for example: (1) Military oriented solutions for SDR equipment are rather costly for disaster applications, (2) Waveform processing in SDR need significant energy and computing resources that is a problem for handheld terminals.

2.4 Cognitive Radio (CR)

Public safety agencies are increasingly using wireless communication technologies to monitor disaster conditions using video surveillance cameras and sensors. The increasing use has led to congestion [17] in radio frequency channels allocated to the agency. In order to address the aforementioned issue of optimum resource allocation during emergency response, CR technology could be exploited to replace the current state-of-the-art channel allocation protocols with an adaptive CE [18].

2.5 Indoor Position Technologies

In disaster scenarios, Global Navigation Satellite Systems (GNSS) based positioning is used to enhance the coordination of the rescue teams. However, due to the lack of GNSS coverage in indoor environments (such as tunnels and buildings), indoor navigation is required for providing the location services to first time responders. In order to make indoor positioning a potential technology for disaster scenario, some of the issues that need to be addressed are: (1) indoor-positioning devices should not be cumbersome to enable their easy deployment, (2) designing energy efficient algorithms for indoor positioning to maximize the battery lifetime of the mobile nodes in a localization system for disaster scenario [19].

2.6 Disaster Situation Aware Protocols for Mobile Devices

The integration of context-aware computing with mobile devices enable them to adapt and react to dynamic changes in the environment. This concept is used in [20] to design a context-aware ad hoc network for effective crowd disaster mitigation by issuing an alert to prevent a stampede in the crowded area. The authors designed Disaster Aware Protocols (DAP) taking into account the disaster situation, allowing mobile devices to be effective in a disaster scenario. In the absence (or partial presence) of an infrastructure, a mobile device should be able to operate in a disaster mode serving as a lifeline for the common people on the ground. DAP for mobile devices should feature communication mode switching scheme in which information such as the amount of remaining battery, mobility (mobile phones movement), and a number of neighboring mobile devices could be used by the mobile phone to decide the apt communication mode [21].

2.7 Mobile Phone Disaster Mode

Mobile phone is a potential device in the event of a disaster scenario to be able to help us connect with family and friends, locate resources, navigate to a safer location and help others. In addition, smartphones can use their integrated sensors to help allocate scarce resources to the most affected people by collecting data to enable disaster relief teams to comprehend the unraveling situation in the disaster zone. However, due to the challenges of energy-management and connectivity, currently available smart phones are not well equipped to operate efficiently during disaster. Often, disaster victims are left helpless with poor or no connectivity.

3 Existing IoT Standards: LoRa/4G LTE

According to ITU, IoT is defined as: A global infrastructure for the information society enabling advanced services by interconnecting things based on, existing and evolving, interoperable information and communication technologies.' [22]. Maximizing the communication of hardware objects and converting the harvested data into a meaningful information without any human involvement, are the two major objectives of IoT. IoT is the combination of three basic elements: hardware, middleware, and presentation [23]. Hardware is further divided into embedded sensors, actuators, and communication systems. The embedded sensors collect data from the monitoring area and send it to the middleware element. Middleware element processes a huge amount of received data and extracts interpretable information with the help of different data analysis tools. Visualization of processed data in an easily readable form gets transformed through the presentation element. Presentation element also processes user queries to the middleware element for necessary actions. Figure 1 shows the block diagram of an IoT system, where different communication standards have been used (in the literature) to communicate between blocks. We will discuss two major standards: LoRa and 4G LTE.

3.1 LoRa

In 2012, Semtech acquired a spread spectrum technique named LoRa. LoRa can be formed by taking the derivative of Chirp Spread Spectrum (CSS). Any MAC layer

Fig. 1 IoT's three main elements and their communication

could be used with LoRa physical layer. However, the currently proposed MAC is Low Range Wide Area Network (LoRaWAN) which works on the principle of simple star topology. LoRa supports a star topology; therefore, it can transmit over a very long distance. Gateway, which is connected to a backbone infrastructure, is directly connected with the nodes. These gateways are powerful devices capable of receiving and decoding a number of concurrent transmissions (up to 50) through powerful radios. Node devices are classified into three classes: (1) Class A end devices: Transmission from node to the gateway only occurs when needed. After transmission receive window is activated to obtain the queued messages through gateway (2) Class B end-devices with scheduled receive slots: Class B operates on a similar principle as Class A node, but with additional receive window (3) Class C end-devices with maximal receives lots: These nodes are not suitable for batter-powered operations due to continuous listening.

3.1.1 Limitations of LoRaWAN

The two main pillars of the IoT growth are transportation and logistics. The efficiency of multiple applications is targeted in areas such as disaster management, public and goods transportation. However, some applications are resistant to swing, delay and fluctuations, while others are not. Delay constraints are diverse for different applications but LoRaWAN being a low power wide area network (LPWAN) solution, is not well-suited for these applications. Contrarily, LoRaWAN supports solutions such as fleet management and control. Similarly, for video surveillance, MJPEG, MPEG-4, and H.264 are the most commonly used digital video formats for IP-based video systems. The data rate recommended for IP surveillance cameras ranges from 130 kb/s with low-quality MJPEG coding to 4 Mb/s for 1920 × 1080 resolution and 30 fps MPEG-4/H.264 coding [24]. The data rate of LoRaWAN ranges from 0.3 to 50 kb/s per channel, thus it is not well-suited for these applications.

3.2 4G LTE

4G LTE is ideal for IoT application not only for its flat all-inclusive nature of IP architecture but also because it has built-in security along with robust and scalable traffic management capabilities. The spectral-efficiency of LTE is greater than second generation (2G) and third generation (3G) networks; therefore, data transmission could be done at a much lower rate. In this regard, data transmission is 2–3 times less costly than 3G while 20 times less than 2G. IoT friendly LTE chipsets are the foundation for the new wave of LTE device development, which is flexible, efficient, and low cost. Numerous LTE chipsets and modules are available today. These innovative solutions provide all required features to build a robust and long life LTE devices for numerous applications at a low cost. Features includes a small footprint and ultra-low power consumption.

3.2.1 Limitations of 4G LTE

4G systems are mainly designed to deal with Human-type Communication (HTC) traffic. Consequently, when considering 5G systems, IoT dictate to simultaneously handle the presence of HTC and Machine-type Communication (MTC) traffic, while meeting the requirements of these traffic types. As a further step, the disruptive technologies are aiming at introducing flexibility, customization, and re-congurability of the network in both radio and core segments, in order to enable the provisioning of enhanced IoT services. Indeed, a natural evolution of connecting devices to the Internet is to remotely control these devices through the Internet. However, for large number of users, 4G LTE suffers from high delay and high packet loss. The work presented in [25] revealed that the efficiency of 4G LTE decreases dramatically as the number of traffic increases.

3.3 Research Recommendations/Future Directions

Indeed, the next generation IoT, WSNs, 5G, and big data analytics stands as a major enabler to realise future intelligent infrastructures for enhanced disaster management. The widespread demand for data and the emergence of new services are inevitably leading to the so-called Resource Crisis. Hence, the evolution of the current centralised model of networked systems to new paradigms such as low power high data rate cognitive networks present a suitable path to counteract this crisis.

The existing LoRa provides transmission parameters such as transmission power, coding rate, spreading factor, and bandwidth, resulting in over 936 possible combinations. These configuration parameters could be optimally tuned to acquire optimized bit-rate, airtime, and energy consumption, taking into account the local electromagnetic environment, constraints, and objectives. For example, increasing spreading factor to improve link reliability nearly halves the datarate and doubles the energy consumption. Similarly, increasing bandwidth doubles the datarate and halves the energy consumption and airtime, reducing link reliability due to additional unwanted noise.

Recent research on LoRa/LoRaWAN has mainly focused on LoRa performance evaluation in terms of coverage, capacity, scalability and lifetime [26–28]. Furthermore, recent work has also proposed adaptive approaches to allocate optimal transmission parameters [29]. However, most of these methods are based on state-of-the-art mathematical/statistical models and suffer from limited modelling assumptions, limited learning, inability to deal with non-linear complex behaviours, poor scalability, and no time-series/temporal data exploitation. Future research should focus on developing robust fair data rate allocation and power control methods to address existing LoRa limitations and acquire optimised airtime, datarate, and energy consumption.

4 Conclusion

The emergence of new wireless communication services and demand for Big Data processing in real-time poses new architectural and radio resource management challenges. This chapter surveys research on emerging wireless communication technologies for effective disaster monitoring and management systems. WSN and IoT stand as a major enabler for enhanced disaster monitoring and management systems. In this chapter, limitations of two major IoT standards (LoRA and 4G LTE) are presented with some future research recommendations. It is concluded that future research should focus on developing artificial intelligence/machine learning driven more robust radio resource management strategies to enable optimized operations in real-time.

Acknowledgements This work was supported by the UK Engineering and Physical Sciences Research Council (EPSRC) Grant No. EP/M026981/1.

References

1. Atzori L, Iera A, Morabito G (2010) The internet of things: a survey. Comput Netw 54 (15):2787–2805
2. Akyildiz IF, Su W, Sankarasubramaniam Y, Cayirci E (2002) Wireless sensor networks: a survey. Comput Netw 38(4):393–422
3. Alemdar H, Ersoy C (2010) Wireless sensor networks for healthcare: a survey. Comput Netw 54(15):2688–2710
4. Chen D, Liu Z, Wang L, Dou M, Chen J, Li H (2013) Natural disaster monitoring with wireless sensor networks: a case study of data-intensive applications upon low-cost scalable systems. Mobile Netw Appl 18(5):651–663
5. Asimakopoulou E, Bessis N (2011) Buildings and crowds: forming smart cities for more effective disaster management. In: Innovative mobile and internet services in ubiquitous computing (IMIS), 2011 fifth international conference on, IEEE, pp 229–234
6. Shah SIA, Fayed M, Dhodhi M, Mouftah HT (2011) Aqua-net: a flexible architectural framework for water management based on wireless sensor networks. In: Electrical and computer engineering (CCECE), 2011 24th Canadian conference on, IEEE. pp 000481–000484
7. Kung H-Y, Hua J-S, Chen C-T (2006) Drought forecast model and framework using wireless sensor networks. Journal of information science and engineering 22(4):751–769
8. Wang P, Sun Z, Vuran MC, Al-Rodhaan MA, Al-Dhelaan AM, Akyildiz IF (2011) On network connectivity of wireless sensor networks for sandstorm monitoring. Comput Netw 55 (5):1150–1157
9. Varnes DJ (1984) Landslide hazard zonation: a review of principles and practice. Number 3

10. Kotta HZ, Rantelobo K, Tena S, Klau G (2011) Wireless sensor network for landslide monitoring in nusa tenggara timur. TELKOMNIKA Telecommun Comput Electron Control 9 (1):9–18
11. Ramesh MV (2014) Design, development, and deployment of a wireless sensor network for detection of landslides. Ad Hoc Netw 13:2–18
12. Terzis A, Anandarajah A, Moore K, Wang I (2006) Slip surface localization in wireless sensor networks for landslide prediction. In: Proceedings of the 5th international conference on information processing in sensor networks, ACM, pp 109–116
13. Lee H-C, Ke K-H, Fang Y-M, Lee B-J, Chan T-C (2017) Open-source wireless sensor system for long-term monitoring of slope movement. IEEE Trans Instrum Meas 66(4):767–776
14. Rosi A, Berti M, Bicocchi N, Castelli G, Corsini A, Mamei M, Zambonelli F (2011) Landslide monitoring with sensor networks: experiences and lessons learnt from a real-world deployment. Int J Sens Netw 10(3):111–122
15. Giorgetti A, Lucchi M, Tavelli E, Barla M, Gigli G, Casagli N, Chiani M, Dardari D (2016) A robust wireless sensor network for landslide risk analysis: system design, deployment, and field testing. IEEE Sens J 16(16):6374–6386
16. Rawat P, Haddad M, Altman E (2015) Towards efficient disaster management: 5 g and device to device communication. In: Information and communication technologies for disaster management (ICT-DM), 2015 2nd international conference on, IEEE, pp 79–87
17. Doumi TL (2006) Spectrum considerations for public safety in the united states. IEEE Commun Mag 44(1):30–37
18. Gorcin A, Arslan H (2008) Public safety and emergency case communications: opportunities from the aspect of cognitive radio. In: New frontiers in dynamic spectrum access Networks, 2008. DySPAN 2008. 3rd IEEE symposium on, pp 1–10
19. Robles JJ, Tromer S, Quiroga M, Lehnert R (2010) A low-power scheme for localization in wireless sensor networks. In: Meeting of the European Network of Universities and companies in information and communication engineering, pp 259–262. Springer, Berlin
20. Ramesh MV, Shanmughan A, Prabha R (2014) Context aware ad hoc network for mitigation of crowd disasters. Ad Hoc Netw 18:55–70
21. Nishiyama H, Ito M, Kato N (2014) Relay-by-smartphone: realizing multihop device-to-device communications. IEEE Commun Mag 52(4):56–65
22. Rose K, Eldridge S, Chapin L (2015) The internet of things: an overview. The internet society (ISOC), pp 1–50
23. Kaur N, Sood SK (2017) An energy-efficient architecture for the internet of things (iot). IEEE Syst J 11(2):796–805
24. Adelantado F, Vilajosana X, Tuset-Peiro P, Martinez B, Melia-Segui J, Watteyne T (2017) Understanding the limits of lorawan. IEEE Commun Mag 55(9):34–40
25. Lloret J, Parra L, Taha M, Tomás J (2017) An architecture and protocol for smart continuous ehealth monitoring using 5g. Comput Netw 129:340–351
26. Oliveira R, Guardalben L, Sargento S (2017) Long range communications in urban and rural environments. In: Computers and communications (ISCC), 2017 IEEE symposium on, pp 810–817
27. Petäjäjärvi J, Mikhaylov K, Pettissalo M, Janhunen J, Iinatti J (2017) Performance of a low-power wide-area network based on lora technology: Doppler robustness, scalability, and coverage. Int J Distrib Sens Netw 13(3):1550147717699412
28. Hosseinzadeh S, Larijani H, Curtis K, Wixted A, Amini A (2017) Empirical propagation performance evaluation of lora for indoor environment. In: Industrial informatics (INDIN), 2017 IEEE 15th international conference on, pp 26–31
29. Bor M, Roedig U (2017) Lora transmission parameter selection. In: Proceedings of the 13th IEEE international conference on distributed computing in sensor systems (DCOSS), Ottawa, ON, Canada, pp 5–7

Modelling of Earthquake Hazard and Secondary Effects for Loss Assessment in Marmara (Turkey)

Ilya Sianko, Reyes Garcia, Zuhal Ozdemir, Iman Hajirasouliha and Kypros Pilakoutas

Abstract This study proposes an innovative Earthquake Risk Assessment (ERA) framework to calculate seismic hazard maps in regions where limited seismo-tectonic information exists. The tool calculates the seismic hazard using a probabilistic seismic hazard analysis (PSHA) based on a Monte-Carlo approach, which generates synthetic earthquake catalogues by randomizing key hazard parameters in a controlled manner. All the available data was transferred to GIS format and the results are evaluated to obtain a hazard maps that consider site amplification, liquefaction susceptibility and landslide hazard. The effectiveness of the PSHA methodology is demonstrated by carrying out the hazard analysis of Marmara region (Turkey), for which benchmark maps already exist. The results show that the hazard maps for Marmara region compare well with previous PSHA studies and with the National Building Code map. The proposed method is particularly suitable for generating hazard maps in developing countries, where data is not available or easily accessible.

Keywords Earthquake · Seismic hazard · Secondary hazard · Liquefaction
Marmara region

I. Sianko · R. Garcia (✉) · Z. Ozdemir · I. Hajirasouliha · K. Pilakoutas
Department of Civil and Structural Eng, University of Sheffield, Sheffield, UK
e-mail: r.garcia@sheffield.ac.uk

I. Sianko
e-mail: isianko1@sheffield.ac.uk

Z. Ozdemir
e-mail: z.ozdemir@sheffield.ac.uk

I. Hajirasouliha
e-mail: i.hajirasouliha@sheffield.ac.uk

K. Pilakoutas
e-mail: k.pilakoutas@sheffield.ac.uk

T. S. Durrani et al. (eds.), *Geological Disaster Monitoring Based on Sensor Networks*, Springer Natural Hazards, https://doi.org/10.1007/978-981-13-0992-2_6

1 Introduction

Earthquakes and their secondary effects represent a major threat to communities, particularly in developing countries where many structures are substandard due to inadequate construction quality and use of poor quality materials. The benefits of enforcing modern seismic codes will be only reflected in future structures. Economic losses and causalities following a future earthquake can be substantially reduced by developing a better understanding of earthquake risks, assessing them before the earthquake happens and implementing appropriate mitigation strategies.

In developed countries, seismic risk assessment is done by specially developed generic tools such as Global Earthquake Model (GEM-Europe), RISK-UE (EU) [15], HAZUS (USA) [10] as well as derivatives such as RADIUS (Japan) [18]. These tools are usually applicable to structures in developed countries and even though they contain useful elements, they are not easily applicable to regions where limited input data exist. For instance, many developing countries lack of detailed instrumental or historical earthquake catalogues, as well as high-quality geological/structural data, which in turn prevents an accurate prediction of risk. To do seismic risk assessment in such countries, simple and effective methodologies relying on limited data are still required.

One of the most critical components in seismic risk assessment is the calculation of hazard. Most of the frameworks developed to date use the Probabilistic Seismic Hazard assessment (PSHA) procedure proposed by Cornell [7], which gives the solution of the total probability theorem for all seismic sources. The main drawback with the application of conventional PSHA is the ground-motion variability. This variability has a large spatial (inter-event or site-to-site) component, and it is implicitly assumed to be entirely temporal (intra-event or earthquake-to-earthquake), thus overestimating the seismic demand on the exposed building stock. More recently, Monte-Carlo (MC) approaches have been also proposed as an alternative to Cornell's approach. Unlike conventional PSHA, MC simulations can take into account more complicated factors such as spatial correlation of ground shaking [1]. Monte-Carlo generates synthetic catalogues to represent seismicity of the area. Due to the fact that the location and time of the earthquake are random, controlled random numbers can generate random events with time and space variations. Random values are generated from predefined distributions to determine the occurrence of earthquakes for every seismic source zone for each year of a catalogue. The location and depth of the epicentre is also determined randomly within the source zone. The magnitude value of the generated event is calculated randomly with reference to the magnitude-frequency distribution for that particular source zone. Each synthetic catalogue represents a scenario of what could occur in terms of earthquakes in a region in the next years, that will be consistent with past behaviour. For each generated earthquake, the ground shaking at the site can be also simulated from knowledge of the attenuation and the scatter of the attenuation [16].

Previous research at the University of Sheffield has developed an Earthquake Risk Assessment (ERA) framework for regions where limited seismo-tectonic information exists (Khan [13] for Pakistan, Kythreoti [14] for Cyprus). However, the applicability of the framework still needs to be extended to other seismic-prone regions. Further improvements to make the calculation of hazard more efficient are also necessary.

This work study proposes an innovative Earthquake Risk Assessment (ERA) framework to calculate seismic hazard maps in regions where limited seismo-tectonic information exists. A stochastic probabilistic seismic hazard analysis tool based on Poisson and time-dependent hazard models is developed by generating synthetic earthquake catalogues using a Monte-Carlo approach. A case study area (Marmara, Turkey) is selected to validate the effectiveness of the tool. The proposed ERA framework also assesses secondary hazards such as liquefaction and landslides. This study is the initial part of a RCUK-TUBITAK Newton Fund project which aims to develop an innovative and rapid framework for multi-hazard seismic risk assessment and disaster mitigation for substandard buildings in Turkey. Post-earthquake disaster intelligence from real-time data from smart devices will be also coupled to the framework. This will contribute to the coordination of post-earthquake search and rescue actions, enhance situational awareness and to validate earthquake risk assessment predictions.

2 Methodology

2.1 Geological and Tectonic Setting

The Marmara region is chosen for verification of the framework. Marmara is one of the most seismically active zones in the world and has produced many large earthquakes with strike-slip faulting. The North Anatolian strike-slip fault zone (NAFZ) extends across northern Turkey for more than 1500 km, and moves about 25 mm/year in right-lateral slip between Anatolia and Eurasian plate [21]. The western part of the NAFZ has a greatest impact on the tectonic regime of the Marmara Sea area: the NAFZ continues as a single fault line east of 31.5°E, whereas to the west is divided into a complex fault system (Fig. 1). Both Kocaeli (August 17, 1999) and Duzce (November 12, 1999) earthquakes were the last events representing propagation of seismic activity towards the west of the NAFZ [3].

Previous PSHA studies for Marmara assuming that earthquakes are occurring independently by utilising a Poisson model [9]. However, the paleo-seismic and historical earthquake evidence from the North Anatolian fault line show that the major characteristic earthquakes on these faults occur with some time interval [19]. By taking into account this time pattern for faults generating earthquakes, it is deemed acceptable for modelling time-dependency (Renewal Method) to represent seismicity of the region.

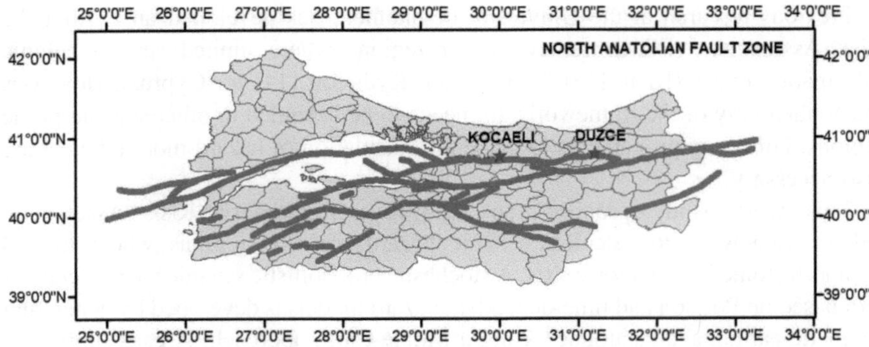

Fig. 1 Faults system in Marmara region with epicentral location of major earthquakes occurred in 1999

2.2 PSHA Input Parameters

The seismic hazard assessment has been carried out mostly following the probabilistic methodology proposed by Musson [16] using the original computer code developed based on proposed methodology for ERA framework. The method consists of definition of background and fault seismic source zones and generating of synthetic catalogues. For background zones, earthquake magnitude-recurrence relationship is applied, while for faults zones characteristic magnitude approach is used. Ground motion prediction equations (GMPEs) uncertainty addressed with a logic tree approach. The methodology is used to produce hazard maps in terms of ground motion and an associated annual frequency of exceedance.

The area for the PSHA is bounded by 39.0°–42.3°N, 25.5°–33.1°E. The background seismicity is assumed to include all earthquakes of $5.5 < M_w < 7.0$. All earthquakes larger than $M_w = 7$ are assumed to occur on faults through characteristic earthquakes, and the fault segmentation model proposed by Erdik et al. (2004) has been used to model the location of these characteristic earthquakes (Fig. 2). The AFAD catalogue has been used to model the historical and instrumental seismicity of Marmara. The catalogue includes events with $M_w \geq 4.0$. The data of earthquakes of $M_w \geq 5.5$ are assumed to be complete from 1900. The Gutenberg–Richter recurrence relationship has been calculated using maximum likelihood regression for the sources of background seismicity using the earthquake events of magnitude $5.5 < M_w < 7.0$ from 1900 until 2016 from the earthquake catalogue. The b-value of the Gutenberg–Richter relationship was calculated for the whole area due to a lack of data in some zones, and it was found to be 0.81. This b value was then set for each zone, whereas the a-value was calculated based on the activity rate of the background source zone.

A background zones source model can be used to construct large numbers of synthetic earthquake catalogues, which represent different possible outcomes of the

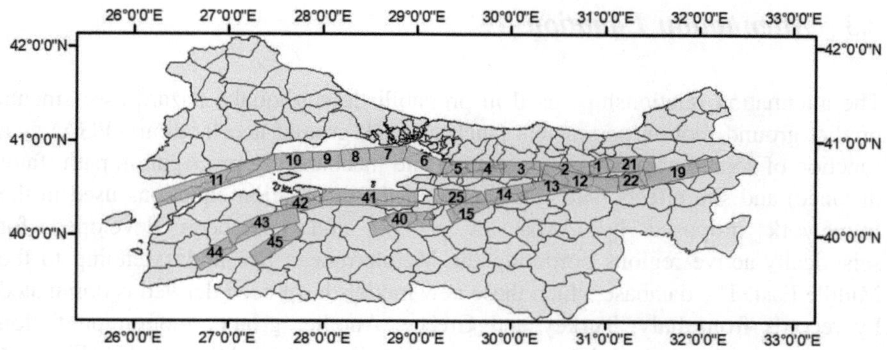

Fig. 2 Location of study area showing boundaries of the provinces in Marmara region, and the fault segmentation model proposed by Erdik et al. (2004)

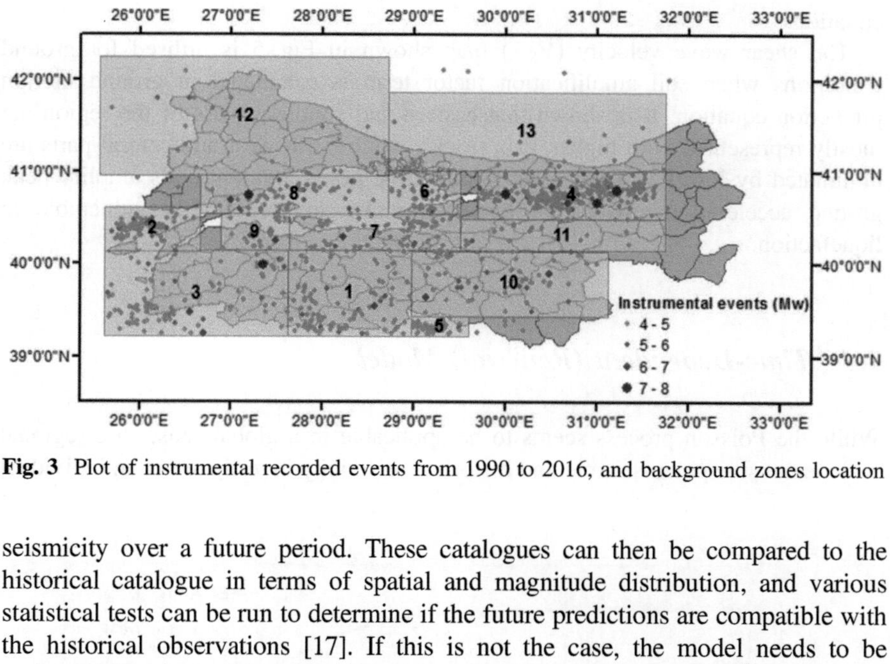

Fig. 3 Plot of instrumental recorded events from 1990 to 2016, and background zones location

seismicity over a future period. These catalogues can then be compared to the historical catalogue in terms of spatial and magnitude distribution, and various statistical tests can be run to determine if the future predictions are compatible with the historical observations [17]. If this is not the case, the model needs to be reviewed closely to determine the source of the discrepancy. The X^2 testing proposed by Musson and Winter [17] is applied to determine background zones properties and location (Fig. 3). It is shown that Marmara is subdivided into 13 background zones, with smaller zone's sizes for higher concentration of events and larger zones for more spread out events.

2.3 Attenuation Equation

The attenuation relationships used in probabilistic earthquake hazard assessments predict ground motion parameters (such as peak ground acceleration—PGA) as a function of source parameters (magnitude and mechanism), propagation path (fault distance) and site effects (site class). One of the attenuation equations used in the framework proposed by Akkar et al. [2] and has been developed for seismically-active regions bordering the Mediterranean Sea and extending to the Middle East. The database which these new models have been derived is dominated by records from Italy, Turkey and Greece. Another ground motion prediction equation used in methodology is proposed by Boore and Atkinson [5]. Figure 4 compares the above GMPEs for $M_w = 7.5$ for rock soil conditions. The plot shows that, between distances 1.0 and 15.0 km, the attenuation equation developed by Akkar et al. [2] predicts lower PGA values. A logic tree approach is used to take into account uncertainty in GPMEs, with a weight factor of 0.5 for each attenuation equation.

The shear wave velocity (V_{S30}) map shown in Fig. 5 is utilised for ground conditions when soil amplification factor term is calculated in ground motion prediction equation. It is shown that eastern and southern parts of the region are mostly represented with higher V_{S30} (rock), while north-west and central parts are dominated by low V_{S30} (soft soil). It should be noted that soft soils amplify peak ground accelerations, which in turn can make an area highly vulnerable to liquefaction.

2.4 Time-Dependent (Renewal) Model

While the Poisson process seems to be applicable in a global sense in a regional scale, extensive paleo-seismic and historical seismicity investigations on individual

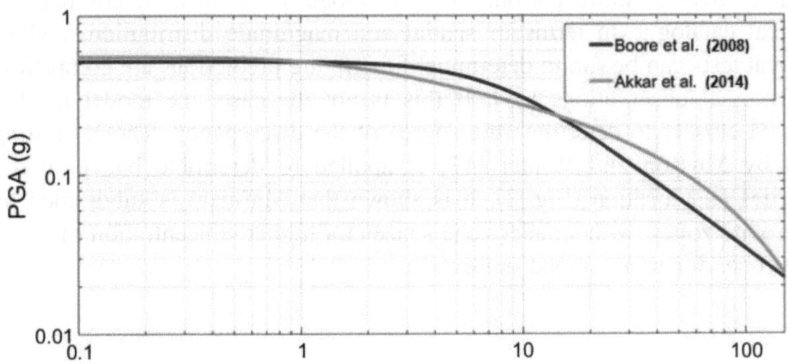

Fig. 4 Example of different GMPEs for $M_w = 7.5$ in terms of PGA and distance

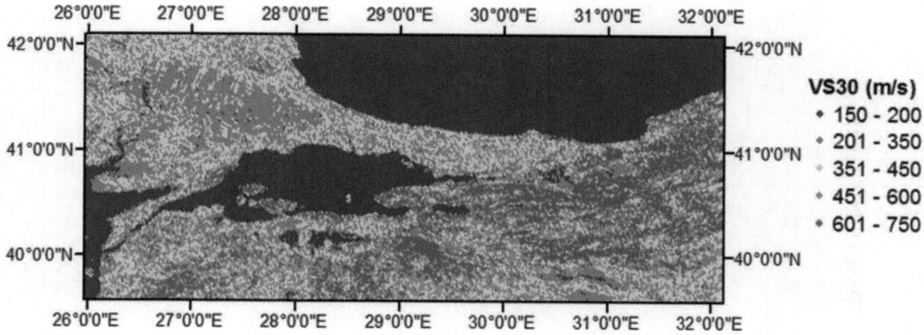

Fig. 5 Shear wave velocity (V_{S30}) map of Marmara region (data from USGS)

faults indicate a somewhat periodic occurrence of large (characteristic) magnitude earthquakes that necessitate the use of 'time dependent' (or 'renewal') stochastic models [19]. The time dependent model assumes that the occurrence of large (characteristic) earthquakes has some periodicity. The conditional probability that an earthquake occurs in the next ΔT years, given that it has not occurred in the last T years is given by:

$$P(T, \Delta T) = \frac{\int_T^{T+\Delta T} f(t)dt}{\int_T^{\infty} f(t)dt} \tag{1}$$

where f(t) is the probability density function for the earthquake recurrence intervals, T is the elapsed period of time since the last major event and ΔT is the exposure period, taken as 50 years (usually taken as life span of the building). An example of probability density function for event with mean return period of 100 years is drawn in Fig. 6. It can be observed that probability is rapidly increasing during the first 100 years, while probability almost not changing after 300 years since last event. Therefore, probability of occurrence will be almost at maximum without increasing significantly for the next years, until the event will occur.

Various statistical models have been proposed for the computation of the probability density function, such as Gaussian, log-normal, Weibull, Gamma and Brownian [9]. In this study, the recently proposed Brownian Passage Time (BPT) model is assumed to adequately represent the earthquake distribution Ellsworth et al. [8].

$$PDF = \left(\frac{\mu}{2\pi\pi^2 t^3}\right)^{1/2} exp\left(-\frac{(t-\mu)^2}{2\alpha^2 \mu t}\right)^{1/2} \tag{2}$$

For the renewal model, the conditional probabilities for each fault segment are calculated. These probabilities are said to be conditional since they change as a

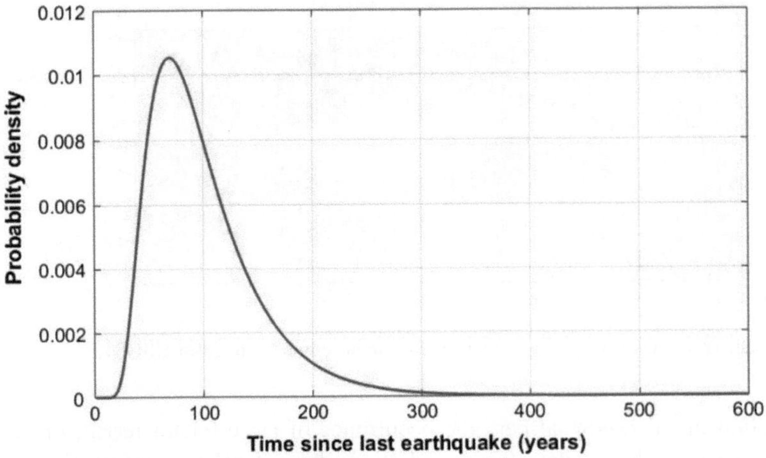

Fig. 6 Probability density functions for Brownian Passage Time distribution with mean 100 years

function of the time elapsed since the last earthquake. A lognormal distribution with a covariance $\alpha = 0.5$ is assumed to represent the earthquake probability density distribution. The calculated 50 year conditional probabilities are converted into effective Poissonian annual probabilities by the use of the following expression [22]:

$$R_{eff} = -ln(1 - P_{cond})/T \tag{3}$$

Table 1 summarises various parameters assigned to each fault segment of the model in the region, including Poissonian and Time-dependent annual rates and Characteristic Magnitude.

Table 1 Poisson and renewal model characteristic earthquake parameters for segments in the model show on Fig. 2 adopted from Erdik et al. [9]

Segment number	Char. magnitude (Mw)	Poissonian annual rate	Mean recurrence time (years)	Time since last earthquake (years)	Time-dependent annual rate
1	7.2	0.0071	140	17	0.0020
2	7.2	0.0071	140	17	0.0020
3	7.2	0.0071	140	17	0.0020
4	7.2	0.0071	140	17	0.0020
5	7.2	0.0057	175	125	0.0102
6	7.2	0.0048	210	265	0.0104
7	7.2	0.0040	250	253	0.0082
8	7.2	0.0040	250	253	0.0082

(continued)

Table 1 (continued)

Segment number	Char. magnitude (Mw)	Poissonian annual rate	Mean recurrence time (years)	Time since last earthquake (years)	Time-dependent annual rate
9	7.2	0.0050	200	463	0.0114
10	7.2	0.0050	200	1000	0.0110
11	7.5	0.0067	150	107	0.0121
12	7.2	0.0040	250	52	0.0010
13	7.2	0.0017	600	1000	0.0037
14	7.2	0.0017	600	1000	0.0037
15	7.2	0.0010	1000	1000	0.0020
19	7.5	0.0040	250	75	0.0022
21	7.2	0.0040	250	20	0.0001
22	7.2	0.0040	250	62	0.0015
25	7.5	0.0010	1000	1000	0.0020
40	7.2	0.0010	1000	164	0.0000
41	7.2	0.0010	1000	1000	0.0020
42	6.8	0.0010	1000	1000	0.0020
43	7.2	0.0010	1000	282	0.0002
44	7.2	0.0010	1000	1000	0.0020
45	7.2	0.0010	1000	66	0.0000

3 Results and Discussion

To develop regional hazard maps, it is essential to quantify seismic hazard associated with a certain ground condition ('reference ground') from which the ground motion for other types of ground condition can be inferred. Figures 7 and 8 show the reference ground characterized with V_{S30} = 760 m/s for a 475 years return period, assuming a Poisson or time-dependent model. For comparison, Figs. 9, 10 and 11 show the seismic hazard map (in terms of PGA for the same return period) obtained from the European Project SHARE (www.share-eu.org/), Kalkan et al. [12], and Disaster and Emergency Management Presidency of Turkey (AFAD, http://www.afad.gov.tr). It is shown that the PSHA results compare well with previous PSHA studies for selected return period. In particular, the results are consistent with the PSHA map developed by AFAD.

Table 2 compares PGA results for specific sites. As can be seen, the proposed methodology matches well the values proposed by AFAD and Kalkan et al. [12]. However, it is also evident that the methodology also tends to predict slightly higher PGA values for Izmit, in particular compared to the values by Kalkan et al. [12]. The differences can be attributed to the numerous parameters involved in the calculation of hazard, which are different for every model.

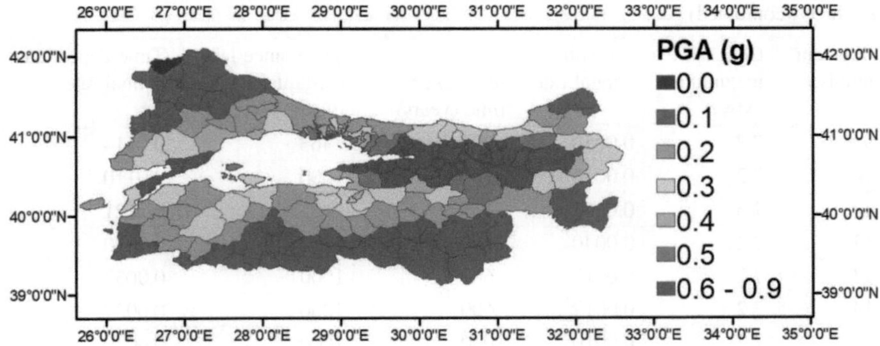

Fig. 7 Poisson model seismic hazard map in terms of peak ground acceleration for the 475 years return period

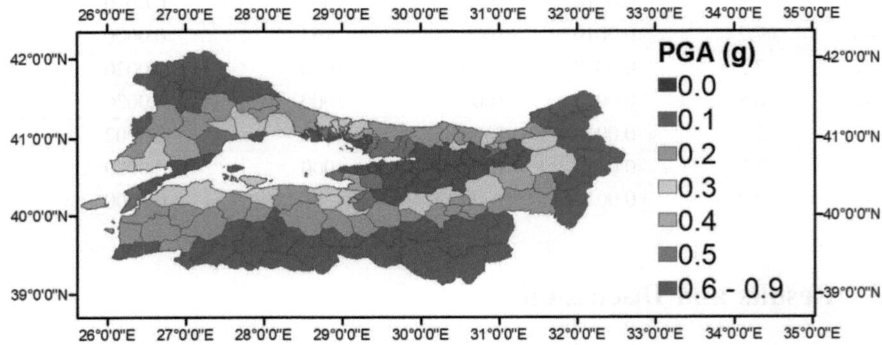

Fig. 8 Time-dependent model seismic hazard map in terms of peak ground acceleration for the 475 years return period

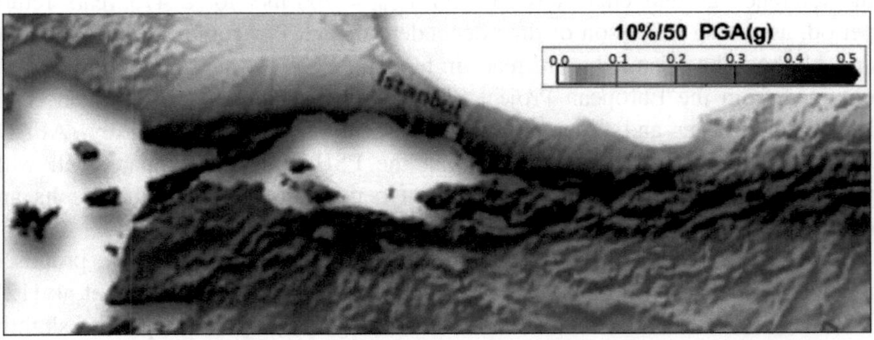

Fig. 9 Seismic hazard map in terms of PGA for 475 years return period, produced by the European Project SHARE (www.share-eu.org/)

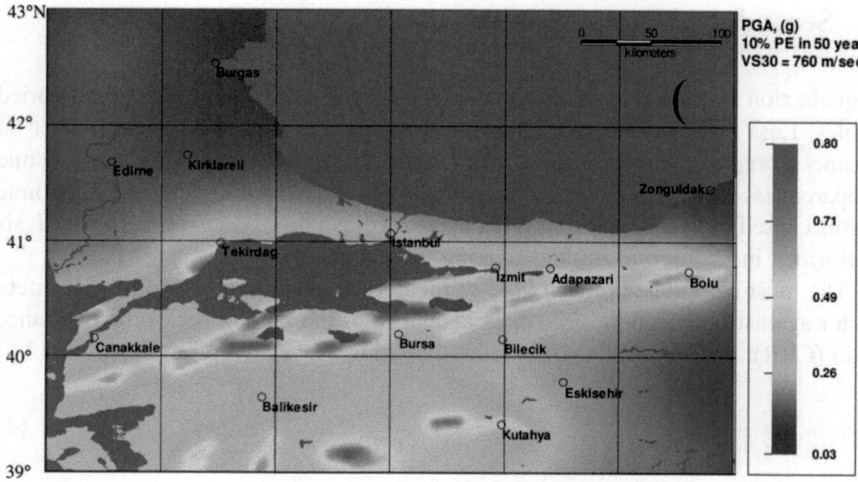

Fig. 10 Seismic hazard map in terms of PGA for 475 years return period produced by Kalkan et al. [12]

Fig. 11 Seismic hazard map in terms of PGA for 475 years return period, produced by AFAD

Table 2 Comparison of PGA values at different location, 475 years return period

Location	This study			
	Time-dependent (g)	Poisson (g)	AFAD (g)	Kalkan et al. [12] (g)
Istanbul	0.37	0.29	0.32	0.31
Izmit	0.75	0.84	0.72	0.51
Bursa	0.39	0.31	0.35	0.32
Tekirdag	0.47	0.38	0.41	0.46

4 Secondary Effects—Liquefaction

Liquefaction can make big damage to lifelines such as roads, pipelines and buried cables. Loss of power and reduction in transport operation are having impact on business organizations running their normal operations. Evaluating the seismic preparedness of infrastructure is necessary to understanding indirect economic losses caused by business interruption and to achieve this, liquefaction risk analysis performed in addition to ground shaking prediction.

The most common approach used to predict liquefaction triggering is the safety factor against liquefaction, *FS*, which is defined as the ratio of the cyclic resistance ratio (CRR), and the cyclic stress ratio, CSR, for a layer of soil at depth, z:

$$FS^{*} = 1.4\frac{CRR}{CSR} \tag{4}$$

According to methodology proposed by Seed and Idriss [20], CSR can be expressed by:

$$CSR = 0.65\left(\frac{a_{max}}{g}\right)\left(\frac{\sigma_v}{\sigma'_v}\right)r_d \tag{5}$$

where a_{max} is the peak horizontal ground acceleration; g is the acceleration of gravity; σ_v is the total overburden stress at depth z; σ'_v is the effective overburden stress at depth z; and r_d is a shear stress reduction coefficient.

CRR is normally determined from geotechnical parameters based on cone penetration test (CPT) or standard penetration test (SPT) results. However, Andrus and Stokoe [4] propose a different approach for calculating CRR based on the shear-wave velocity:

$$CRR = \left[0.022\left(\frac{V_{S1}}{100}\right)^2 + 2.8\left(\frac{1}{V^*_{S1} - V_{S1}} - \frac{1}{V^*_{S1}}\right)\right] \times MSF \tag{6}$$

where V_{S1} is the stress-corrected shear wave velocity; V^*_{S1} is the limiting upper value of V_{S1} for cyclic liquefaction occurrence, which varies between 200 and 215 m/s depending on the fines content of the soil; and MSF is a magnitude scaling factor.

The calculated value for V_{S10} can then be used as a proxy for V_S at all soil layers between 0 and 10 m depth and both the V_{S10} and V_{S20} values can be used to determine an equivalent proxy for all soil layers between 10 and 20 m. From manipulation of the Boore et al. [6] empirical functions and the formula for calculating averaged shear wave velocities, the following equations determine the proxies to be used in the two depth ranges:

$$V_{S(0-10)} = 10^{\left(\frac{\log V_{S30} - 0.042062}{1.0292}\right)} \tag{7}$$

$$V_{S(10-20)} = \frac{1}{\frac{2}{10^{\left(\frac{\log V_{S30} - 0.025439}{1.0095}\right)}} - \frac{1}{V_{S(0-10)}}} \tag{8}$$

The liquefaction potential index (*LPI*), which predicts the possibility of liquefaction at surface-level by integrating a function of the factors of safety for each soil layer within the top 20 m of soil. *LPI* is calculated as:

$$LPI = \int_0^{20} F^*(10 - 0.5z)dz \tag{9}$$

where $F^* = 1 - FS^*$ for a single soil layer. The soil profile can be sub-divided into any number of layers [11] calibrated the *LPI* model and determined guideline criteria for determining liquefaction risk.

According to this criterion, liquefaction risk is very low for $LPI = 0$; low for $0 < LPI \leq 5$; high for $5 < LPI \leq 15$; and very high for $LPI > 15$.

Figure 12 shows the results of applying Eqs. 4–9 to the case study area. The results indicate that liquefaction potential is high for places with soft soil, while very low risk exists for places located predominantly on rock. It should be noted that due to lack of accurate data regarding the water table in Marmara, the results in Fig. 12 cannot be considered as definitive. In this study, the water table and soil densities across the region were assumed as equal to 10 m across the whole region, dry soil density $\gamma = 10$ kN/m^3, and saturated soil density $\gamma = 20$ kN/m^3. Future research will need to update the liquefaction hazard map using real data and compare them with existing studies for selected sites.

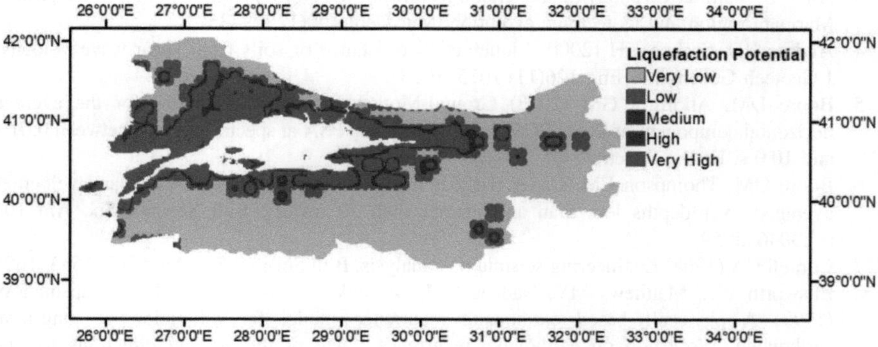

Fig. 12 Initial liquefaction map of Marmara region based on 10% in 50 years' seismic hazard

5 Conclusions

A Probabilistic Seismic Hazard (PSHA) methodology is proposed for implementing into a new Earthquake Risk Assessment framework for regions where limited seismo-tectonic data exist. This PHSA method utilises a Monte-Carlo approach to build synthetic earthquake catalogues that are generated by randomizing the key hazard parameters of earthquake magnitude, epicentral location, depth of hypocentre, and basic tectonic and geological parameters. The method is demonstrated by carrying out a PSHA study for Marmara, Turkey. The seismic hazard results compare well with previous PSHA studies. In particular, the results are consistent with the PSHA map included in the most recent National Building Code of Turkey. This suggests that the method described can be acceptable for producing hazard maps in regions where limited data are available. The proposed framework can be extended to other developing regions around the world and its results can be used to assist relevant stakeholders and decision-makers on preparedness, emergency response and mitigation actions.

Acknowledgements The research leading to these results has received funding from the RCUK-TUBITAK Research Partnerships Newton Fund Awards under grant agreement EP/P010016/1.

References

1. Akkar S, Cheng Y (2016) Application of a Monte-Carlo simulation approach for the probabilistic assessment of seismic hazard for geographically distributed portfolio. Earthq Eng Struct Dyn 45(4):525–541
2. Akkar S, Sandıkkaya M, Bommer J (2014) Empirical ground-motion models for point-and extended-source crustal earthquake scenarios in Europe and the Middle East. Off Publ Eur Assoc Earthq Eng 12(1):359–387
3. Alpar B, Yaltırak C (2002) Characteristic features of the North Anatolian Fault in the eastern Marmara region and its tectonic evolution. Mar Geol 190(1):329–350
4. Andrus RD, Stokoe KH (2000) Liquefaction resistance of soils from shear-wave velocity. J Geotech Geoenviron Eng 126(11):1015–1025
5. Boore DM, Atkinson GM (2008) Ground-Motion Prediction equations for the average horizontal component of PGA, PGV, and 5%-damped PSA at spectral periods between 0.01 s and 10.0 s. Earthq Spectra 24(1):99–138
6. Boore DM, Thompson EM, Cadet H (2011) Regional correlations of VS30 and velocities averaged over depths less than and greater than 30 meters. Bull Seismol Soc Am 101 (6):3046–3059
7. Cornell CA (1968) Engineering seismic risk analysis. Bull Seismol Soc Am 58(5):1583–1606
8. Ellsworth WL, Matthews MV, Nadeau RM, Nishenko SP, Reasenberg PA, Simpson RW (1999) A physically-based earthquake recurrence model for estimation of long-term earthquake. Workshop on earthquake recurrence. State of the art and directions for the future, Istituto Nazionale de Geofisica, Rome, Italy, 22–25
9. Erdik M, Demircioglu M, Sesetyan K, Durukal E, Siyahi B (2004) Earthquake hazard in Marmara Region, Turkey. Soil Dyn Earthq Eng 24(8):605–631

10. FEMA (1999) Earthquake loss estimation methodology technical manual
11. Iwasaki T, Arakawa T, Tokida KI (1984) Simplified procedures for assessing soil liquefaction during earthquakes. Soil Dyn Earthq Eng 3(1):49–58
12. Kalkan E, Gülkan P, Öztürk NY, Çelebı M (2008) Seismic hazard in the Istanbul metropolitan area: a preliminary re-evaluation. J Earthq Eng 12:151–164
13. Khan SA (2011) An earthquake risk assessment framework for developing countries: Pakistan a case study. PhD thesis
14. Kythreoti S (2002) Earthquake risk assessment and management. Case study: Cyprus. PhD Thesis
15. Mouroux P, Bertrand E, Bour M, Brun Bl, Depinoise S, Masure P, RISK-UE (2004) The European Risk-Ue Project: an advanced approach to earthquake risk scenarios. In: 13th World conference on earthquake engineering, Vancouver, B.C., Canada
16. Musson R (1999) Determination of design earthquakes in seismic hazard analysis through Monte Carlo simulation. J Earthq Eng 3(4):463–474
17. Musson R, Winter P (2012) Objective assessment of source models for seismic hazard studies: with a worked example from UK data. Bull Earthq Eng 10(2):367–378
18. Okazaki K, Villacis C, Cardona C, Kaneko F, Shaw R, Sun J, Masure P, Mouroux P, Martin C, Davidson R, Tobin LT (2000) RADIUS, Risk assessment tools for diagnosis of urban areas against seismic disasters, UN
19. Schwartz DP, Coppersmith KJ (1984) Fault behavior and characteristic earthquakes—examples from the wasatch and San-Andreas Fault zones. J Geophys Res 89(NB7):5681–5698
20. Seed HB, Idriss IM (1971) Simplified procedure for evaluating soil liquefaction potential. J Soil Mech Found Div 97(9):1249–1273
21. Straub C, Kahle HG, Schindler C (1997) GPS and geologic estimates of the tectonic activity in the Marmara Sea region, NW Anatolia. J Geophys Res: Solid Earth 102(B12):27587–27601
22. WGCEP94 (1995) Seismic hazards in Southern California: probable earthquakes, 1994 to 2024. Bull Seismol Soc Am 85:379–439

Unmanned Aerial Vehicles for Disaster Management

Chunbo Luo, Wang Miao, Hanif Ullah, Sally McClean, Gerard Parr and Geyong Min

Abstract This chapter highlights the communication and network technologies that contribute to UAV disaster management systems, surveys the latest development of UAV-assisted disaster management applications, including early warning system, search and rescue, data gathering, emergency communication, and logistics, and presents our preliminary work to demonstrate the benefits and challenges of UAV systems for emergency communication. Finally, we discuss the characteristics and design challenges of UAV disaster management systems.

Keywords Unmanned aerial vehicle · UAV · Disaster management systems
Network · Communication

1 Introduction

Large-scale natural disasters test the most fundamental human instinct of survival by inflicting massive and often unpredictable loss of life and property. Different types of natural disasters, such as climatological (extreme temperature, drought, and wildfire), meteorological (tropical storm, hurricane, sandstorm, and heavy rain-fall), hydrological (flash-floods, debris flow, and floods), and geophysical (earthquake, tsunami, volcano, landslide, and avalanche) have resulted many deaths and huge economic damage [1]. According to a report published by Münchener Rückversicherungs-Gesellschaft (Munich R. n.d.) in 2017, the tsunami which struck in 2004 was the deadliest natural disaster occurring worldwide in the last four decades. The disaster claimed almost 222,000 lives. Similarly, the earthquake

C. Luo (✉) · W. Miao · G. Min
University of Exeter, Exeter, UK
e-mail: c.luo@exeter.ac.uk

H. Ullah · S. McClean
University of Ulster, Coleraine, UK

G. Parr
University of East Anglia, Norwich, UK

T. S. Durrani et al. (eds.), *Geological Disaster Monitoring Based on Sensor Networks*, Springer Natural Hazards, https://doi.org/10.1007/978-981-13-0992-2_7

that affected Haiti in January 2010, was the second deadliest disaster that cost approximately 159,000 lives. Also, according to another survey conducted by the United States Geological Survey (USGS), nearly 316,000 people died, 30 million were injured, and 1.3 million were displaced [2]. In addition to the fatalities mentioned earlier, billions of U.S. dollars were invested in the rehabilitation process resulting in huge economic damage. According to another statistic, the earthquake and tsunami in Japan in 2011 caused an economic damage of almost 210 billion U. S. dollars [3]. To predict and identify the incidences of natural disasters in order to respond on a timely manner and to reduce the number of fatalities and economic damage, there is therefore a need for an enhanced and efficient network-assisted disaster management system [4].

Unmanned Aerial Vehicles (UAV) have shown their power to tackle some of the most challenging tasks to human beings, such as natural disasters. To develop and deploy UAV-based solutions has attracted great interest from both academia and industry worldwide. Although natural disasters cannot be avoided, their effects could be significantly mitigated through a comprehensive, and highly efficient disaster management system. Time is the most essential factor for the success of disaster management. How to quickly assess the degree of the disaster effects, and design reasonable disaster response to address urgent and most severe scenarios is a challenging issue.

Traditional disaster management systems that rely only on ground emergency responses suffer from the issues of low efficiency in information gathering and disaster assessment, slow identification of victims, and limited communication services including large latency and low bandwidth satellite communication. Further challenging issues that disaster rescuers face are logistics. Following a severe disaster event, streets and roads of the affected area could be blocked by debris or destructed, how to quickly access the disaster area and victims becomes the most critical factor for a successful disaster management. One typical disaster is, for instance, the 7.8 magnitude earthquake in Nepal, which claimed the lives of 9000 and injured 23,000 people [5].

Significant research efforts have been dedicated to reduce the adverse effects of natural disasters towards the human society and environment. UAVs have been introduced to disaster management to enhance the capability of disaster responses, deal with the extreme environmental conditions and conduct timely rescue work [6]. During a disaster event, UAVs could bring rescuers the bird's-eye view of the disaster environment, which is very important for a comprehensive disaster management system that involves data collection, victim localization and rescue optimization.

UAVs can be embedded into the entire life-cycle of disaster management. Police departments, fire brigades, and disaster response departments will have access to medium and small UAVs and will be able to integrate UAVs into their rescue work. At the pre-disaster stage, UAVs can be used together with the existing early warning systems to accurately predict the outbreak time and scope, and reduce the cost of economy and materials that disaster may cause through disaster prevention plan. During the disaster, UAVs can provide high-resolution real-time images of even the inaccessible locations, which can then be used to quickly produce accurate hazard maps to guide the rescuer to assess the situation, make relief plans and conduct rescue. And after disaster, UAVs can be used to map the affected areas in high resolution within a short time, which, in turn, can aid swift and efficient response.

2 UAV Networking Technologies

Networking technologies are essential to the success of UAV deployment in practical environments. In this section, we focus on the networking technologies of UAV systems for disaster management and response. First, we introduce a universal network architecture that integrates 802.11 Wi-Fi network, 2G/3G/4G cellular network, UAV ad hoc network and satellite network [7]. Key networking components will be discussed and analysed in detail with an emphasis on determining whether the existing technologies could meet the performance of UAV disaster management.

2.1 Network Architecture

A universal network architecture for UAV-based disaster management is illustrated in Fig. 1, which consists of Wi-Fi network, 2G/3G/4G cellular network, UAV ad hoc network and GPS assisted satellite network.

At the top layer, UAVs connect to the GPS satellite network by equipping a GPS receiver on-board, which periodically provides the geolocation and time information. The information is critically important for UAVs to accurately and safely accomplish the disaster response tasks. Examples include the Google Loon project: balloons deployed by Google at the high altitude around 18 km above the ground, proving cellular and internet services for remote and rural areas [8]. For UAV-based disaster management, the high-altitude balloons provide effective means to establish communication among the UAVs, and between the UAVs and ground stations in a large-scale geographic area. Another example is the EU ABSOLUTE project, which creates a hybrid kite-balloon, named Helikite, to rapidly deploy 4G communications for emergencies [9]. Two types of network connectivity exist in the envisioned network architecture, named infrastructure-based connectivity and ad hoc network connectivity.

Infrastructure-based connectivity

For infrastructure-based network connectivity, each ground node, e.g. LTE eNode base station or WI-FI access point, can easily communicate with aerial UAVs, and thus establish a star topology network structure where the UAV is at the center of the star. Under this topology, the ground station can indirectly communicate with

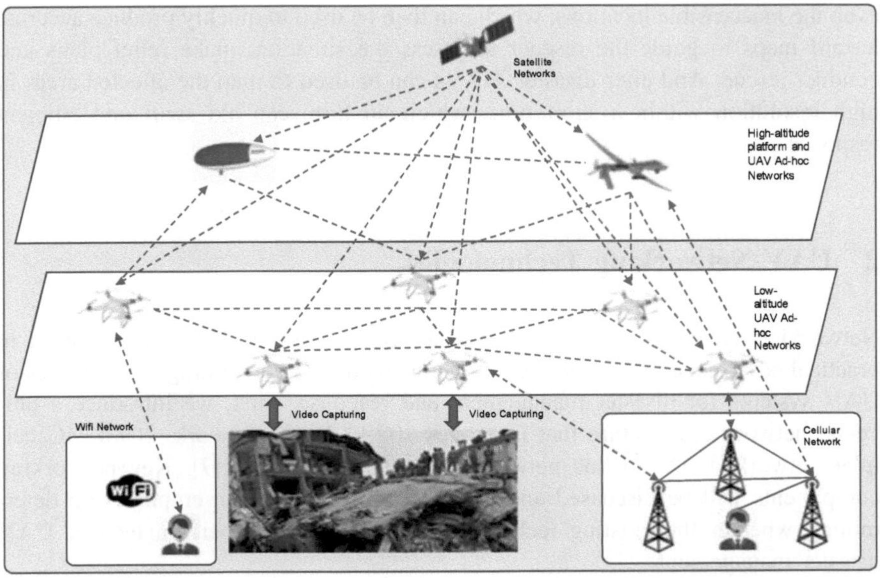

Fig. 1 A universal network architecture for UAV-based disaster management

others over the UAV. Infrastructure-based network connectivity has the benefits of easy deployment, configuration, and reliable channel conditions, while it suffers from limited coverage and bandwidth limitation when high-definition videos are required to transmit in the large area for disaster analysis and recovery. One solution to this issue is to deploy a sophisticated UAV at high-altitude, covering a large area, collecting valuable data, storing it on-board and discharge it later to the ground station when UAVs return back, or exchanging information with the ground eNode stations in real time.

Ad hoc-based network connectivity

Multiple low-cost mini UAVs form an aerial ad hoc network, as shown in Fig. 1. For disaster management, mini UAVs could fly at lower altitudes and accomplish tasks in a distributed manner. They do not require specific launching device, and can take-off and land at various types of terrains, making them very suitable for disaster environment. In addition, during a strong disaster event, a team with mini-UAVs are capable to satisfy on-demand service request at fine-grain level. However, a lot of challenging issues exist when multiple UAVs are practically deployed in disaster events. For instance, mini-UAVs have restricted capabilities such as power, sensing, communication, and computation. They cannot fly for a long time in the air and are hard to maintain stable and reliable communication links. Therefore, more research efforts are needed to achieve reliable and high-performance ad hoc-based UAV network connectivity.

It is worth mentioning that a UAV disaster management system should adaptively choose the most suitable type of connectivity, depending on the disaster scenarios and the network infrastructure available in the affected area. If the 2G/3G/4G cellular networks are working normally, then UAVs can be connected to the base stations. Otherwise, ad hoc connectivity may be formed to transmit the information captured by UAVs to ground control stations in a multi-hop way. In the next section, we will introduce the system components and performance requirements of different UAV applications.

2.2 System Components

The holistic system architecture of UAV networks (shown in Fig. 1) is composed of three main components: a ground control centre or operator, ground and aerial communication networks, and UAV devices.

- The ground control centre is responsible for the overall control of the whole UAV system, flight task design, data storage and analysis as well as decision making in case of emergency. It provides the interface for human users to interact with the UAVs in the air.
- Communication networks provide reliable and high-performance communication for a UAV system to transmit and receive the information to/from the ground control centre. As shown in Fig. 1, various wireless communication approaches have been exploited to provide ubiquitous communication access for disaster management. Their characteristics and performance are summarized in Table 1.
- UAV devices provide the essential platform for sensing and communication, e.g. data gathering, processing, and transmission. In addition, it may also run various algorithms to maintain safety and tackle emergencies such as battery exhaustion and signal loss.

UAVs play a key role in the universal network architecture. Therefore, the capacity, performance, and operation theatre of the UAV system directly relate to its abilities, such as endurance, range, altitude, and payload [10, 11]. To use UAVs in different applications, an emerging requirement of decreasing the personnel needs and increasing the autonomy of UAVs, by which they can fly freely in the sky and carry out the mission without any centralized controller from the ground station, is becoming more and more clear. UAVs used in disaster management should be pilotless aircrafts that do not require direct human intervention for flying and operating. They should navigate autonomously according to the pre-programmed software and can be controlled remotely.

Table 1 Characteristics of wireless communication technologies in disaster management

Feature	Wi-Fi	GPS	UMTS	LTE	LTE-A
Channel width	20 MHz	2 MHz	5 MHz	1.4, 3, 5, 10, 15, 20 MHz	Up to 100 MHz
Frequency band	2.4 GHz, 5.2 GHz	1176–1576 MHz	700–2600 MHz	700–2690 MHz	450 MHz–4.99 GHz
Bit rate	6–54 Mb/s	50 b/s	2 Mb/s	Up to 300 Mb/s	Up to 1 Gb/s
Range	Up to 100 m	\	Up to 10 km	Up to 30 km	UP to 30 km
Coverage	Intermittent	Ubiquitous	Ubiquitous	Ubiquitous	Ubiquitous
Mobility support	Low	Extremely high	High	Very high (350 km/h)	Very high (350 km/h)
QoS support	Enhanced distributed channel access	None	QoS classes and bearer selection	QCI and bearer selection	QCI and bearer selection
Broadcast/multicast support	Broadcast	Broadcast	MBMS	eMBMS	eMBMS
UAV support	Yes	Yes	Potential	Potential	Potential
Market penetration	High	High	High	Potentially high	Potentially high

3 UAV for Emergency Communication

Cooperative wireless networks composed of small unmanned aerial vehicles (UAVs) are easy and fast to deploy, flexible, and inexpensive and can provide on-the-fly communication facilities through self-managed ad hoc Wi-Fi networks to coordinate the rescue teams in case of disastrous events and to help the survivors in a timely manner [12]. Figure 2 shows a complete architecture of a UAV-based cooperative wireless network deployed over a large geographical area in order to provide on-the-fly communication facilities in a disaster affected area. In such situations, UAVs are used as communication relays to bridge communication in those areas where part of the communication infrastructure is destroyed due to natural disasters such as earthquakes, flood or man-made disasters such as bomb-blast, military attacks etc.

3.1 UAV as a Communication Relay

UAVs as communication relays have got much attention in recent years and a lot of work have been done in this area. In [13], we proposed an optimal UAV deployment algorithm (OUDA) in order to quickly deploy the UAV to an optimal position to bridge communication between nodes on the ground and to provide the best communication facilities to the participating ground stations. The algorithms work in a way that the UAV will start flying toward the disaster area and will begin

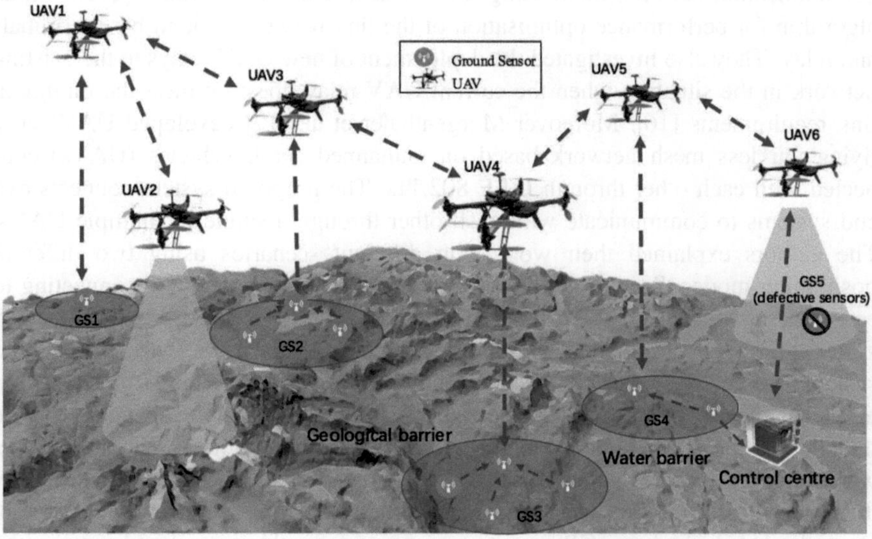

Fig. 2 UAV-based cooperative wireless network for disaster management

sending beacon messages at regular intervals. Once the participating nodes get the beacon message, in response the nodes send their ID along with the GPS position back to the UAV. The UAV will also obtain the received signal strength (RSS), and distance between the UAV and participating nodes on the ground. A mathematical model was also developed for our proposed approach and based on that mathematical model the optimal position of the UAV was calculated with respect to transmitted signal and distance between UAV and nodes. An extensive simulation was carried out to verify the applicability of the proposed algorithm for real-time applications in the context of disaster management.

Similarly, in [14], the authors analyzed the deployment of a UAV as a flying base station. They used the idea to provide on-the-fly wireless communication facilities to a specific geographical area for a determined device-to-device communication network. The authors explained their problem considering two different scenarios with a static UAV and a mobile UAV. The authors continue to say that the overall communication rate and coverage could be improved significantly if the UAV is precisely moved over the specified area [14].

Moreover, Jagun and Hales in [15] presented a dynamic positioning algorithm in order to bridge communication between participating nodes using a UAV as a relay. The authors claim that this approach could be used in real-life situations such as an earthquake and collecting data from distributed sensor nodes. The approach works in a way that a direct link can be established between nodes that are in the range of the UAV. The authors also proposed a scheduling framework where they prioritised different nodes based on the frequency of visit and communication range in order to cover the entire set of distributed nodes [15].

Furthermore, in [16], the authors explored a communication system with some ground based terminals along with a network base station with a view to bridging communication between them using a UAV as a relay. The authors developed an algorithm for performance optimisation of the link between ground base terminals and relay. They also investigated the deployment of new UAV relays to the existing network in the situation when the current UAV relay does not meet the minimum link requirements [16]. Moreover Morgenthaler et al. [17] developed UAVNet, a flying wireless mesh network based on unmanned aerial vehicles (UAVs) connected with each other through IEEE 802.11s. The proposed system connects two end systems to communicate with each other through a single or multiple UAVs. The authors explained their work with different scenarios using two different positioning modes. Scenario one was carried out with a single UAV connecting to end systems located on the ground. The UAV starts flying towards the first end system and begins to transmit ping messages on a regular basis. Once the first end system received the message, it sends back its GPS position to the UAV. Based on either manual searching mode, where the first end system will tell the UAV the direction of the second end system, or based on the autonomous searching mode, where the UAV will calculate series of waypoints on a spiral track and will follow this route until it receives the GPS location message from the second end system. Once the UAV obtains a GPS position message from the second end system, the

UAV controller will calculates the central position of the two end systems based on the submitted GPS positions or signal strength and will deploy the UAV at that particular position [17].

3.2 *Bridging Communication Through Multiple UAVs*

Complex situations occur when a single UAV is unable to cover a disaster area in a specific mission. Multiple UAVs are used in such cases in order to cover a set of nodes in a bounded region. A cooperative distributed planning algorithm termed as the consensus based bundle algorithm (CBBA) was proposed in [18] in order to ensure network connectivity with limited communication facilities for a group of heterogeneous agents operating in a dynamic environment. The algorithm with UAV as relays improves the mission performance and range of the team without any constraint on the active agents. CBBA used those agents which are free as relays for network connectivity to the base station during task execution. A set of assignments of agent-task pairs is created from the pool of initial tasks so that the process of bidding can be carried out on each task. During this bidding process some of the tasks may not be assigned and that may result in network disconnection. The unassigned tasks that are disconnected from the base are introduced as new relay tasks, and based on their assignments they will create a connected network. The authors validated the algorithm based on simulation results and through indoor and outdoor experimental field tests, and concluded that CBBA is well suited for real time applications [18]. Similarly, swarms of UAVs as communication relays for range and reliability improvement of ground based ad hoc networks were studied in [19]. The authors envisaged the application of distributed multiple-input-multiple-output (MIMO) schemes on multiple UAVs in a cluster, and also noticed the performance of distributed transmit beamforming and distributed orthogonal space-time block codes (OSTBC) techniques for different UAV flights under ideal and non-ideal conditions. Different simulation parameters were used to verify the performance of both the aforementioned techniques. The BER performance along with different Rice K-factors effects for both schemes were examined. The authors concluded that in case of lower carrier frequencies, better performance with distributed beamforming can be achieved. The authors also showed that multi-hop relaying based on a swarm of UAVs could improve the performance of ad hoc networks with respect to range and reliability and could be more useful in tactical scenarios [19].

Complexity increases when a swarm of UAVs are deployed to provide coverage to a group of nodes that are moving in a constrained area. A non-cooperative game technique where each UAV independently decides its future location without the cooperation of any central planning agent was proposed in [20]. The proposed approach used the UAV as a player and the mobile nodes as their payoffs. The UAVs have enough data in order to estimate the locations of other mobile nodes and UAVs and to independently execute their decisions. This kind of approach is more advantageous in the optimization process in a sense that the UAVs have full freedom to come-up with the same solution without the involvement of any central decision-making agent, and also allow the UAVs to complete the mission with full autonomy due to the onboard decision making about the next waypoint. The proposed algorithm also termed as a game comprises of three main elements i.e. set of players, set of strategies for each player, and set of payoffs with respect to selected strategies by each player. In such a game, each player tries to maximize their share of the available resources in order to compete with the co-players. Each player selects different actions from his available set of action also called the strategy, and based on these strategies each player will try to maximize his payoffs. The players need to be restricted to their own strategies because a deviation from the selected strategy will results a reduction in the player's payoff. The set of strategies where the players cannot deviate from his selected strategy is known as Nash Equilibrium (NE). The author highlighted two states of the NE, one is called the Pure Strategy Nash Equilibrium (PSNE), where a single action is taken against each strategy and against each action the best responses from all players are coincided, while the second one is called the Mixed Strategy Nash Equilibrium (MSNE), where the probability distribution law is followed for the range of strategies. The author devised a method for solving n-player games using the concept of MSNE. Based on the simulation results the author concluded that non-cooperative game techniques are more useful for coordinating the movement of communication of UAVs over the area coverage missions [20].

Furthermore, a heuristic algorithm for dynamically placing UAVs as relay nodes for providing communication facilities to disconnected groups of MANET ground nodes was proposed by Chandrashekar et al. [21]. The authors also addressed the issue of mobility for ground nodes and also to update the location of UAVs with respect to the movement of nodes. The algorithm takes the node connectivity matrix generated from distance and terrain constraints, and current location as an input variable in order to deploy the UAV. Clusters of nodes are also detected using the same connectivity matrix and based on this particular information the number of UAVs along with their optimal position for each time frame are calculated. The authors also evaluated a communication framework for facilitating communication between ground nodes and UAVs using the existing MANET protocols. The proposed algorithm was verified based on simulation results and comparative analysis with an exhaustive search algorithm [21].

3.3 Bridging Communication to Connect Disjoint Group of Segments

Heimfarth and Araujo in [22] addressed the issue of connecting disjoint group of wireless sensor networks using a UAV. The UAV works as a data mule that moves across the network to collect data from different disjoint segments. Each segment has their own cluster head that is responsible for any type of communication with UAV. The authors perceived very little packet loss for reasonable traffic along with substantial latency because of the limited speed of the UAV making the system ideal for delay tolerant applications [22]. Similarly, Zhu et al. [23] investigated the problem of deploying a new UAV to an existing UAV deployed network in order to maintain the connectivity of ground mobile ad hoc networks (MANETs). The authors assumed a scenario where the current UAVs were unable to maintain the connectivity of MANETs due to the movement of ground nodes and restricted communication range. The idea of the Minimum Steiner Tree problem with existing mobile Steiner points under edge length bound constraints was used to articulate the problem. The authors used three existing UAV aware approximate algorithms i.e. Deploy-Before-Movement (DBM), Move-Before-Deployment (MBD), and Deploy-Across-Movement (DAM). The authors concluded that DAM algorithm is much better than the other two algorithms in terms of performance for all scenarios and can reduce the deployment of the number of new UAVs up-to 70% [23].

Furthermore Heimfarth and Araujo extended their previous work by prioritizing and visiting those disjoint segments having more incoming/outgoing traffic compared to other segments with low communication traffic [24]. The authors also addressed the issue of UAV movement which was the main contribution of their current article. Two different operating modes with respect to a UAV were studied i.e. search mode and traffic relay mode. In search mode, the UAV just scans the entire network and stores the position of the cluster head for each new disconnected segment, while in packet relay mode, the UAV visits all the cluster heads sequentially in order to get the statistics of communication traffic for each segment. From the simulation results the authors concluded that by prioritizing the segments, having more communication traffic could increase the packet delivery rate up to 70% and could decrease the average latency from 600 s to 240 s. The authors also assert that without prioritizing the highly demanded segments could lead to the mismanagement of resources and should only be used in the case of evenly distributed traffic [24].

Moreover, a cooperative multiple input multiple output (MIMO) technique was used to extend the lifetime of a sensor network and to support communication among the static sensors in a sparse network using a UAV as a relay in order to keep the WSN connected [25]. Two main concepts were combined to sort out the problem of disconnections in the case of scattered WSN: the first one was the use of UAV as a relay to bridge isolated sensor nodes, and secondly the use of MIMO techniques to prolong the communication range of these static sensor nodes. To keep the connectivity alive, the base station periodically sends beacon messages to its neighbour UAVs. Upon receipt, the UAVs forwards the same beacon message to their neighbour UAVs having the identification information of the sender UAV along with its distance from the base station in terms of hops. An algorithm was proposed based on RSSI to stop the UAV movement in a situation when the UAV is on the edge of the communication range. The algorithm identifies that if a UAV has only one neighbour and the RSSI received from the last neighbour indicates that the communication link is near to break-point, the algorithm forces the UAV to move back to its previous position in order to keep the connection alive. Similarly, instead of conventional MIMO arrangement, the authors used the concept of MISO (multiple input single output) where a cluster of nodes sends data to the UAV. In the case of MISO, the transmitting sensors first share the information with each other and then forward it to the receiver at the same time. The authors concluded that introducing MIMO based communication could considerably extend their group range and could ease the UAVs movement constraints [25].

3.4 Real-Time UAV-Assisted Disaster Management

In this section, we will highlight some of the work that is carried out in real-time environments to check the applicability and applications of the UAV-based wireless networks for disaster management. Guillen-Perez et al. in [26] proposed a UAV-based aerial wireless network in order to extend the coverage of wireless systems. The experiments were carried out by using the IEEE 802.11 standard versions such as a/b/g/n/ac. Five dBi omnidirectional antennas were used to extend the coverage. The performance of a Galileo board used as an intermediate node was investigated in both indoor (controlled lab) and outdoor environment. A maximum throughput of 10.5 and 11 Mbps with a packet-size of 512 and 1024 bytes respectively was claimed at infrastructure mode, while a throughput of 4.5 and 7.5 Mbps with the same 512 and 1024 bytes was claimed at ad hoc mode using the IEEE 802.11n at both 2.4 and 5 GHz band. Similarly, the radio coverage for both downlink and uplink frequencies at 2.4 and 5 GHz band using the 802.11a/b/g/n/ac was also calculated in the presence of both Friis and Winner D1 propagation models. The authors pointed that the radio-coverage attained during the real experiments is quite different from the one observed during the theoretical study. The authors concluded that the systems perform better in terms of coverage, throughput, and energy efficiency in infrastructure mode, while in terms of current consumption, the system performs better in ad hoc operational mode [26]. Similarly, in [27] the authors investigated the challenges of wireless communication between UAVs in search and rescue missions where the UAVs equipped with cameras need to deliver high resolution images to rescuers through high speed communication. Real-time testbed experiments were conducted based on IEEE 802.11n protocol to check the quality of UAV-to-UAV links under varying context parameters. The experiments were conducted in an outdoor environment to investigate the effects of distance, speed, and rate adaptation over link quality and network performance. The authors concluded that the throughput achieved during the experiments using the IEEE 802.11n is not only far from the maximum throughput observed during the theoretical setup, but also varies drastically even in the case of constant distance between UAVs [27].

Furthermore, Yanmaz et al. [28] proposed a high-level architecture for the design of a collaborative aerial system consisting of quadcopters in order to demonstrate its potentials in disaster assistance, search and rescue operations, and aerial monitoring. In addition to the UAV platform, the authors also identified some other main building blocks such as sensing, coordination, and communication modules of a multi-UAV system. The main purpose of this research was to provide support to firefighters during disaster, to extend the coverage capabilities with no mission time constraints, and real-time video support for search and rescue operations. The application areas mentioned earlier were explored with respect to different coordination, sensing, and communication constraints such as: for time-critical missions (search and rescue operations) with varying objectives, the authors proposed that distributed coordination, reliable sensing and networking capabilities are required. Real-world tests and experiments were performed to observe the applicability of the proposed multi-UAV systems for dynamic applications [28].

3.5 *Results and Discussion from Real-time Testbed Experiments*

In this section, we will discuss some of the results based on the experiments performed in an outdoor environment through a communication link facilitated by an Access Point (AP) mounted on a UAV between the two participating ground stations and a UAV. The results presented in Fig. 3 are based on the work carried out in [29], while the results presented in Fig. 4 are based on the experiments carried out in [30]. In [29], we proposed a UAV-based wireless network to bridge communication between two ground stations and a UAV having an access point mounted on it using IEEE 802.11n. We investigated the performance of 802.11n at both 2.4 and 5 GHz band in terms of data rate, link quality, received signal strength (RSS), and signal-to-noise ratio (SNR) in order to check the applicability of the proposed network for rescue operations and disaster management scenarios. Figure 3 shows the link quality and signal to noise ratio of 802.11n at 2.4 and 5 GHz band for both the client and server machines. As evident from Fig. 3a, b the link quality is quite good for both client and server machines during the experiment (above 70%); it is worth noting that 802.11n at 2.4 GHz has better link quality as compared to 802.11n at 5 GHz. Similarly, Fig. 3c, d shows the SNR of 802.11n at both 2.4 and 5 GHz for both client and server machines respectively. Again, the SNR is lower in the case of 802.11n at 2.4 GHz, while higher in the case of 802.11n at 5 GHz. Low SNR values represent better signal quality, while high SNR values represent poor signal quality. Interference because of wind in an outdoor environment was basically the reason behind such a high SNR.

Similarly, in [30], we extended the work of [29] and performed some outdoor experiments in order to check the performance of IEEE 802.11a/b/g/n. We analysed the performance of 802.11a/b/g at 2.4 GHz and 802.11n at both 2.4 and 5 GHz band in terms of data rate, signal-to-noise ratio, and received signal strength. Three different scenarios were envisioned in terms of UAV altitude in order to check the

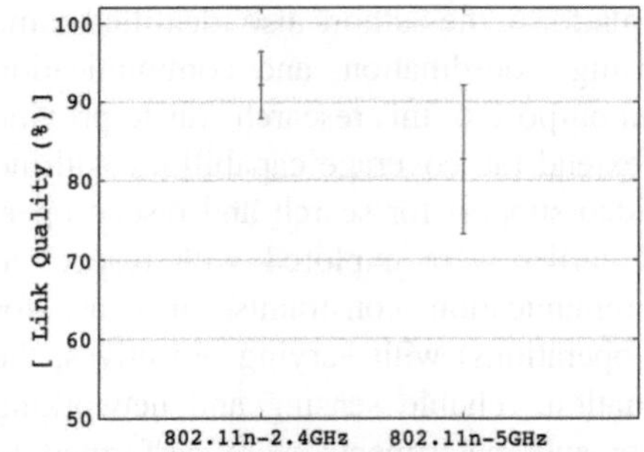

(a) Link quality of 11n at client side

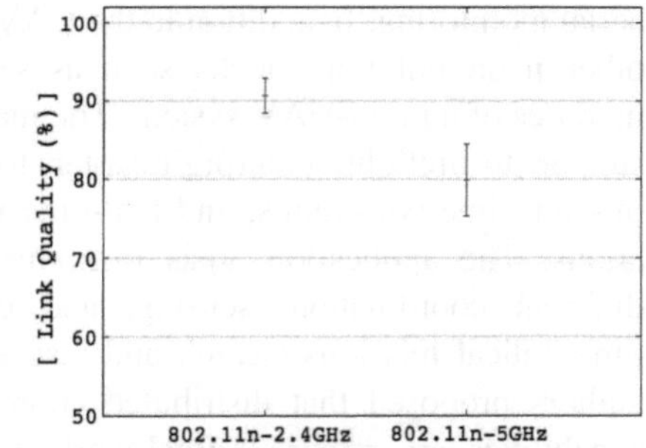

(b) Link quality of 11n at server side

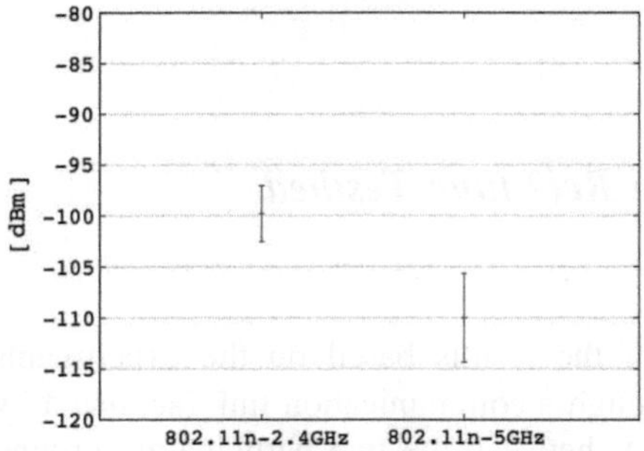

(c) SNR at client side

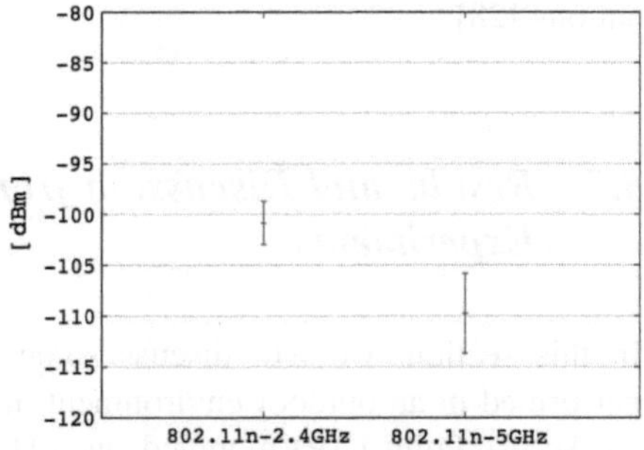

(d) SNR at server side

Fig. 3 Link quality and signal-to-noise ratio of IEEE 802.11n at 2.4 & 5 GHz band

applicability of the proposed systems in disaster management situations. In scenario 1, we calculated the data rate, signal strength, and signal to noise ratio of between UAV and ground stations at 10-meter height, while in scenario 2, and 3, we repeated the same experiments for the same metrics at a height of 15-meter and 20-meter respectively. The ground stations were used as a client and server, where the client send a data of 10 MB to the server through a communication link facilitated by 802.11 protocol stack. Figure 4 shows the data rate of IEEE 802.11a/b/g/n at 2.4 GHz and 802.11n at 5 GHz band using a 20 MHz channel at 10 m (Fig. 4a), 15 m (Fig. 4b), and 20 m (Fig. 4c) respectively. As evident from the figure, the data rate captured by using 802.11a/b/g at 2.4 GHz is very low in all three cases and is not practicable in real-time disaster management events. Instead, the data rate captured by using 802.11n at both 2.4 and 5 GHz band is quite impressive and can be practiced in real-time catastrophic events.

3.6 Summary

Starting with the calamities that hit different parts of the world in the last two decades resulting in millions of fatalities and economic damage in terms of billions

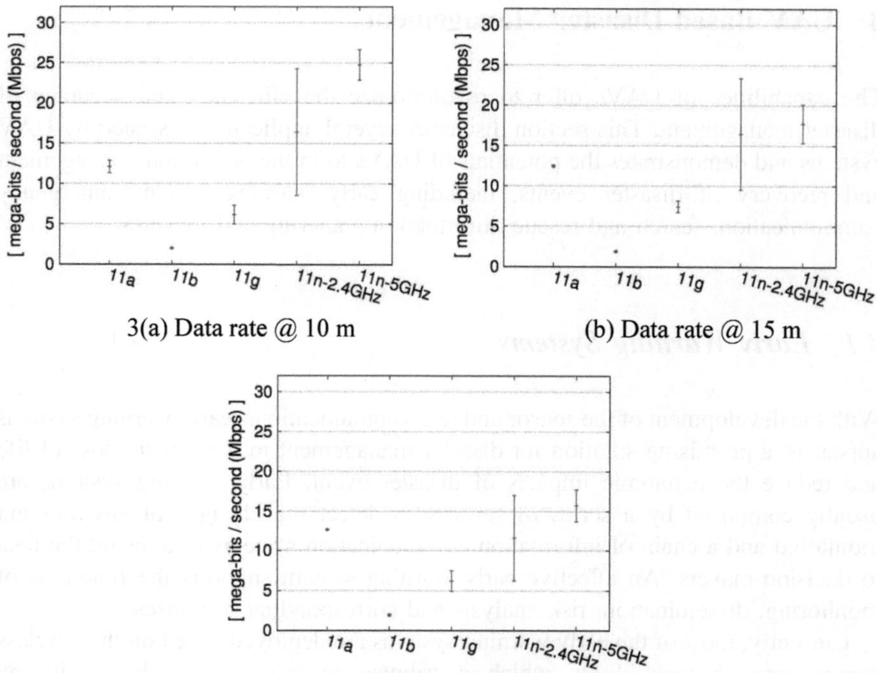

3(a) Data rate @ 10 m

(b) Data rate @ 15 m

(c) Data rate @ 20 m

Fig. 4 Data rate of IEEE 802.11 a/b/g/n-2.4 GHz/n-5 GHz band

of dollars, we highlighted some key areas that could provide help in different catastrophic events and could facilitate timely rescue services to reduce the casualties and economic damage. UAVs as communication relays can provide on-the-fly communication facilities in situations where part of the communication infrastructure is destroyed and there is a need for help on emergency basis to coordinate the rescue teams and help the survivors in a timely manner. Similarly, we investigated some of the key algorithms developed for bridging communication through multiple UAVs in order to extend the coverage area and connect disjoint group of segments. Moreover, we also covered real-time UAV deployment scenarios that could help by providing communication facilities in disaster management situations. Finally, we discussed some of the experimental results from our own work in terms of bridging communication between ground stations and UAV in the context of rescue operations and disastrous events.

4 UAV Based Disaster Management

The capabilities of UAVs offer to revolutionize the efficiency and accuracy of disaster management. This section discusses several applications assisted by UAV systems and demonstrates the potentials of UAVs to in the prediction, management and recovery of disaster events, including early warning system, emergency communication, search and rescue, information gathering and logistics.

4.1 Early Warning Systems

With the development of the sensor and telecommunications, early warning systems appear as a promising solution for disaster management to prevent the loss of life and reduce the economic impacts of disaster event. Early warning systems are usually composed by a series of sensors to detect the changes of environment monitored and a chain of information communication systems to transmit the data to decision-makers. An effective early warning system supports the functions of monitoring, dissemination, risk analysis and corresponding responses.

Currently, most of the early warning systems are deployed based on the wireless sensor network technology, which distributes sensors to a wide monitoring geographic area and interconnects them through wireless links. These sensors are used to monitor physical metrics, such as temperature, moisture, vibration, pressure and so on. Although a huge amount of research work has been dedicated to the performance of sensor-based early warning system, some practical challenges make these early warning systems unsuitable for extreme environment. For instance, the sensor nodes are connected through multiple wireless communication links and the monitoring data need to be sent to sink node through multiple intermediate nodes. In this case, any node failure or broken link would degrade the data delivery. A simple example is that sensor nodes for water monitoring may be washed away in a flood event, compromising the whole warning system. Although, 3G/LTE/LTE-A cellular networks become popular in sensor-based early warning systems to enable data communication, they are not supported by all sensors due to high power consumption.

UAVs are exploited in early warning systems coupled with WSN to provide a comprehensive and high-efficient warning for disaster prediction. There are several ways that UAVs could be used to work together with WSNs in disaster warning.

1. UAVs can be regarded as mobile base stations or data mules in the sky to provide the communication between WSN nodes and decision-making centre. In this case, the data could be uploaded to the aerial UAVs and downloaded by decision-making centre, avoiding the transmission of multiple intermediate nodes, greatly enhancing the reliability and stability of data transmission [31].

2. The second scenario that UAV can be integrated to WSN-based early warning systems is when a certain node or part of network architecture is failing. UAVs could be deployed as relay nodes in the network to deliver data among remote WSN nodes, which is very important for building delay-tolerant networks.
3. The node of WSN network is driven by battery, which makes the energy as a critical important issue in the design, deploy and management of early warning system. Once the power is run out, the node will stop working. In this case, UAV could be exploited as an aerial charger to fly to the exhausted node to refill the battery. There are research publications [32] to investigate the transmitting the power and data simultaneously by exploiting the mobility of UAVs in disaster management. Therefore, UAV have become an important part for designing, deploying and managing early warning system.

4.2 *Emergency Communication*

Emergency communication is an important part of disaster response and recovery. The aim of emergency communication is to provide the communication for the disaster rescuers, affected people, and plan makers. Reliable and easy access communication system is the key to the success of disaster management work, such as information gathering, search and rescue and disaster assessment. As shown in Sect. 3, various communication and information technologies have been exploited in disaster management. The interoperability of the existing wireless communication technologies, such as WiFi, cellular, ad hoc networks as well as aerial base station should be enhanced to improve the capability of wireless communication. Although traditional telecommunication infrastructure could provide high-speed internet and wireless cellular service, they may be disrupted by large-scale nature disasters, for instance the damage to the cable and base station tower due to earthquake, or the shutdown of power supply in a flood event. As a result, the usual mode of communication like mobile devices might not work or might be lost in the calamity.

To address the urgent need of infrastructure independent communication, UAV based solutions have been proposed. The team in [33] designed a UAV-based WiFi system, which largely extends the working distance of the traditional signal from 100 m to 25 km through deploying WiFi module in UAV devices. In addition, directional antenna is used in this system to focus the transmission energy in a certain angle. This work is conducted based on a single UAV device. Multiple UAVs become popular to form an ad hoc network to provide the wireless signal to a wide area. The authors in [34] proposed a multi-UAV enabled wireless communication system, where each UAV carries an aerial base station and provides wireless communication to ground users. To achieve the fairness among the ground users, a maximization-minimization problem is formed by taking UAV's trajectory and power consumption into account and solved by successive convex optimization techniques. To demonstrate the effectiveness of UAVs for post disaster communication, authors of [35] proposed a UAV based emergency communication system, being deployable in large-scale disaster events.

4.3 Search and Rescue

Disaster response is always a race against time, to identify and approach the disaster victims as fast as possible. The primary goal is to save lives. UAVs could assist disaster responders to achieve this goal through quickly scanning large disaster area in the sky and identify potential victims with the help of on-board cameras and navigation sensors. Some interesting research results have been reported in literature. To design an efficient UAV-based search and rescue system, several fundamental factors need to be taken into account, including the energy constricts of UAV systems, quality of the data collected, various hazards of the environment UAV is operating, as well as the QoS of data transmission among UAVs.

The authors in [36] analysed the effects of these factors on the performance of UAV-based search and rescue systems, and studied the optimisation criterions of different search algorithms. Mario et al. [37] proposed a multipurpose UAV for mountain rescue operations. Different environment requirements for mountain search and rescue have been considered in the design of this multipurpose UAV, for instance, the low temperature, high altitude, various payloads and weather conditions. The proposed UAV is equipped with high-performance camera to provide the visual and thermal information. The UAV could be deployed in various search and rescue missions such as identifying victims in heavy snow and woods during night or day. In addition, this multipurpose UAV is designed to be able to carry up to 5 kg payload on-board. Emergency kits or life-support device could be dropped to the victim. The researchers in [38] investigated the effectiveness of UAVs in the search and rescue mission of a hurricane disaster. An emergency response simulation was built to conduct UAV-based search mission in a hypothetical country affected by a hurricane with subsequent flooding and landslides. The test demonstrated that a UAV could only provide limited support for search and rescue, and the authors also pointed out that UAVs have the potential to become part of the emergency response toolkit for very specific tasks. Zheng et al. proposed an integrated search and rescue system by combining aerial and ground unmanned robots [39]. The ground vehicles are responsible for delivering aerial devices in the disaster area and serving as the launching platform for UAVs. Once the UAVs take off from the ground vehicle, they will quickly scan the disaster area and acquire global visual information, which will be used by the ground vehicle to find the optimal path. Therefore, ground and aerial vehicles work cooperatively to accomplish the search and rescue work.

4.4 Information Gathering

Disaster damage assessment becomes essential for rapid relief actions. For decision making, various information is collected and transmitted; therefore, information gathering, or fusion is important for disaster management. For instance, satellite imagery has the advantages to offer visual data of a large-scale geographic disaster area, but these images are usually taken from the space in a vertical perspective, with limited spatial resolution. On the other hand, UAV devices could be deployed to observe the disaster areas in detail, but with limited coverage. Therefore, how to combine the images or video obtained from different platforms and form a unified model of the disaster scenario is a critical and challenging issue for disaster responses.

To address this issue, a novel framework was proposed in literature [40]. Image processing techniques were exploited to combine the advantages of different aerial platforms. The small scale images obtained from UAV devices are used to detect a facade by integrating the geometrical transformation and environment information. Then the large-scale vertical information such as roof and facade features obtained by satellites is used to assess the disaster range. Finally, the overall damage is estimated by information fusion. The authors claimed that their approach could enhance the accuracy of damage estimation in comparison with the roof-only damage estimation approaches. Corrado et al. in [41] investigated how to enhance the performance of an existing UAV device for disasters assessments. The information of radiation detection, camera vision, as well as GPS information is jointly considered. The authors pointed out that the primary challenges for information gathering for low-cost commercial UAVs in disaster assessment is dynamicity of the operation environment, including data transmission range, endurance, payload, and control range. To achieve the integration of multi-sensor data, data fusion models and methods should be developed for a single UAV platform and multiple UAVs [42, 43].

4.5 Logistics

As discussed in the first part of this chapter, one of the most challenging issues in disaster management is logistics. The streets and roads may be blocked or damaged during or after a severe disaster, and it is difficult for the rescue personnel to approach the victims and provide necessary treatments. UAV could be used to drop emergency supplies to victims of disaster areas, such as telecommunication devices, food, and medicine. Current civilian UAVs could only carry up to several kilograms of payload and may not be efficient solutions in logistics for frequent request of emergency drops.

Significant progresses have been achieved to maximize the capability of UAVs in logistics, for instance, fix-wing UAVs are used to transport medicines into remote or inaccessible regions [44]. Compared with the traditional helicopter approaches, UAV-based logistics significantly reduce the cost of delivery and could be affordable for poverty communities. Similar efforts were made by DHL and Amazon to deliver a small quantity of medicine to remote areas. In addition, Google revealed that they also tested UAVs for logistics in the Google X program. John et al. in [45] proposed a coordinated logistics consisted of a truck and a fleet of UAVs. The truck carries the parcels and is automatically operated. The UAVs are launched from the truck with a parcel. Once it is delivered successfully, the UAV will come back to the truck to pick up another one, until all parcels have been delivered. This work investigated how the coordinated system enhances the quality of service for logistics.

5 Design Challenges and Consideration

This section discusses the design challenges and consideration of UAV based ad hoc network for disaster management. Comparing with existing ad hoc based networks, e.g. vehicular ad hoc network (VANET) and mobile ad hoc network (MANET), UAV ad hoc network exhibits different characteristics. The unique features of transmission come from the very specific mission requirements of various aerial applications, e.g. fine-grained oil-pipe or avalanche patrols or flooding disaster response. The specific requirements of the aerial applications pose new challenges for system design, network performance requirement, and communication optimization, considering the factors of delay tolerance, transmission range, topology changes, mobility issue, performance scalability, energy constraints as well as heterogeneous devices.

For example, tight timing synchronization must be met among multiple UAVs when they cooperatively work on the disaster rescue, while large communication delay could be tolerated in the applications such as disaster area scanning. High data rate is expected for sending visual data to identify trapped people, while other applications such as air quality monitoring of the disaster area require low transmission rate. Therefore, different applications have different requirements on the design of the aerial network.

In this section, we first analyse the characteristics of UAV networks in the following three aspects: bandwidth consumption, transmission latency and disruption prone networks. We then discuss the design consideration and challenges with the aim to obtain a complete understanding of UAV system design.

5.1 *Characteristics of UAV Networks*

Bandwidth consumption
For visual data related UAV applications, e.g. search and rescue, videos are captured by on-board cameras and required to be transmitted to the ground control centre timely. They have strict bandwidth requirements, especially for online applications such as real-time monitoring. In addition, the quality of the video captured largely determines the amount of the information that could be used in decision making, therefore, 4K and even 8K videos become increasingly popular in aerial monitoring applications, and push the bandwidth consumption much higher. On the other hand, there are many constraints on the available bandwidth, such as, capacity of the communication channel, speed of UAVs, error-prone structure of the wireless links, lack of security with broadcast communication. The protocols and algorithms designed for the UAV system must satisfy the bandwidth requirement for transmitting high resolution real-time image or video under these constraints.

Transmission latency
For UAV systems, transmission latency is of importance in the design of system components and network protocols. Different aerial applications have different requirements of transmission latency. 3GPP began to specify the latency requirements for aerial applications. For sensed traffic that requires real-time visual or audio data transfer (e.g. research and rescue), the delay should be no more than 50–100 ms. For surveillance, such as border or highway surveillance, the delay can be up to 3 s. Low latency is defined to achieve the online perspective of the aerial commissions, and more important to ensure safe flight, e.g. collision avoidance. The work in [33] investigated the behaviour of packet delay of UAV networks, and concluded that UAV networks exhibit different behaviours than MANETs, and require new aerial transmission protocols and algorithms.

Disruption prone networks
UAV networks are prone to link disruption. The protocols and algorithms should be designed with precautions of dealing with link failure in short or long periods. The disruption level of a link depends on a few factors, including the mobility of UAVs, the distances among them as well as the communication environments such as interference and noise. For the applications that UAVs are hovering over the air, the probability of link disruption will be low. While if the UAVs are operated in high mobility, there exists a high probability of link disruption. Another factor that may impact the quality of the transmission link is the failure of UAV devices. Due to malfunctions or cyber attacks, a UAV may not function properly and fail to communicate with other UAVs. New designs should take disruption into consideration, particularly upper level routing algorithms.

5.2 Design Considerations and Challenges

Channel Modelling: The 3D nature of a UAV network demands the support of various types of channels. The links in an aerial network can be either air-to-air (A2A), air-to-ground (A2G) or ground-to-air (G2A). These links have been analysed against each other as well as against ground-ground (G2G) links [46, 47]. They should be modelled differently according to their distinctive channel characteristics, which affect the supportable network related QoS, and hence the sustainable traffic.

Antenna Structure Design: Antenna structure design in UAV networks should be considered specifically, because link quality fluctuates with antenna orientation. How to efficiently design the structure of UAV antennas to support reliable communication between UAVs and the infrastructure is a timely challenge, with few efforts being demonstrated [48].

UAV Cooperation: Mini-UAVs are small in size and can only carry limited payloads. Facing the rich options such as radars, infrared cameras, thermal cameras, and image sensors, they need to work cooperatively among the team, and with the ground network such as a cloud to effectively complete the tasks in disaster management [49].

Advanced data analytics: One of the advantages of solution-focused algorithms is the ability to apply them across industries, e.g. emergency response, agriculture, energy, even identifying toxic waster in the air in real-time. Advanced algorithms such deep neural networks may provide the much needed functions.

Path Planning: Cooperation and coordination among UAVs are crucial to increase efficiency. In the operation theatre, there can be some dynamic changes like joining/removal of UAVs, physical static obstacles, dynamic threats, etc. In these cases, each UAV has to change its previous path, and new ones should be re-calculated dynamically. High performance algorithms/methods in dynamic path planning are required [50].

Energy efficiency: As UAVs are usually limited by energy, protocols should take special attention on energy efficiency and aim to greener designs. Energy could become a vital factor in disaster management, for example, good design could enable UAVs reach further areas and stay longer for disaster response.

Regulations: While UAVs are gaining their share of the country's national airspace, their deployment has to be regulated in order to maximise the benefits and minimise the potential harms. Regulation may be the biggest barrier for deployment of UAVs in some civilian applications. Properly designed rules and regulations are of practical importance.

References

1. Ruckversicherungs-Gesellschaft M (2015) Loss events worldwide 1980–2014. Geo Risks Research, NatCatSERVICE
2. nd MR (2018) The 10 most significant natural disasters worldwide by death toll from 1980 to 2016. Statista. Available at https://www.statista.com/statistics/268029/natural-disasters-by-death-toll-since-1980/. Accessed 5 Feb 2018
3. nd MR (2018) The 10 biggest natural disasters worldwide by economic damage from 1980 to 2016 (in billion U.S. dollars). Statista. Available at https://www.statista.com/statistics/268126/biggest-natural-disasters-by-economic-damage-since-1980/. Accessed 5 Feb 2018
4. Erdelj M, Natalizio E, Chowdhury KR, Akyildiz IF (2017) Help from the sky: leveraging uavs for disaster management. IEEE Pervasive Comput 16(1):24–32
5. Government of Nepal Ministry of Home Affairs (2015) Nepal disaster report
6. Hein D et al (2017) An integrated rapid mapping systems for disaster management. Int Arch Photogram Remote Sens Spatial Inf Sci vol XLII-1/W1
7. Gupta L et al (2016) Survey of important issues in UAV communication networks. IEEE Commun Surv Tutorials 18(2)
8. Rakesh MT (2016) Study on Google‘s loon project. Int J Adv Res Comput Eng Technol (IJARCET) 5(5)
9. Sathyanarayanan Chandrasekharan et al (2016) Designing and implementing future aerial communication networks. IEEE Commun Mag 54(5)
10. Wang Y et al (2014) Three-dimensional wireless sensor networks: geo-metric approaches for topology and routing design. In: The art of wireless sensor networks, signals and communication technology. Springer, Berlin, p 367–409
11. Luo C, Parr G, McClean SI, Peoples C, Wang X (2015) Hybrid demodulate-forward relay protocol for two-way relay channels. IEEE Trans Wireless Commun 14(8):4328–4341
12. Rosati S, Kruželecki K, Heitz G, Floreano D, Rimoldi B (2016) Dynamic routing for flying ad hoc networks. IEEE Trans Veh Technol 65(3):1690–1700
13. Ullah H, McClean S, Nixon P, Parr G, Luo C (2017) An optimal UAV deployment algorithm for bridging communication. In: 15th international conference on ITS telecommunications (ITST), pp 1–7
14. Mozaffari M, Saad W, Bennis M, Debbah M (2016) Unmanned aerial vehicle with underlaid device-to-device communications: performance and tradeoffs. IEEE Trans Wireless Commun 15(6):3949–3963
15. Jagun K, Hailes S (2014) Scheduling UAVs to bridge communications in delay-tolerant networks using real-time scheduling analysis techniques. In: System integration (SII), 2014 IEEE/SICE international symposium on 2014, pp 363–369
16. Luo C, McClean SI, Parr G, Wang Q, Wang X, Grecos C (2014) A communication model to decouple the path planning and connectivity optimization and support cooperative sensing. IEEE Trans Veh Technol 63(8):3985–3997
17. Morgenthaler S, Braun T, Zhongliang Z, Staub T, Anwander M (2012) UAVNet: A mobile wireless mesh network using unmanned aerial vehicles. In: Globecom Workshops (GC Wkshps), 2012 IEEE, pp 1603–1608
18. Ponda SS, Johnson LB, Kopeikin AN, Han-Lim C, How JP (2012) Distributed planning strategies to ensure network connectivity for dynamic heterogeneous teams. IEEE J Sel Areas Commun 30(5):861–869
19. Palat RC, Annamalau A, Reed JR (2005) Cooperative relaying for ad hoc ground networks using swarm UAVs. IEEE Military communications conference (MILCOM), pp 1588–1594, 17–20 Oct 2005
20. Charlesworth PB (2014) Using non-cooperative games to coordinate communications UAVs. In: Globecom Workshops (GC Wkshps), pp 1463–1468

21. Chandrashekar K, Dekhordi MR, Baras JS (2004) Providing full connectivity in large ad hoc networks by dynamic placement of aerial platforms. IEEE Military communications conference (MILCOM), pp 1429–1436, 31 Oct–3 Nov 2004
22. Heimfarth T, de Araujo JP (2014) Using unmanned aerial vehicle to connect disjoint segments of wireless sensor network. In: IEEE 28th international conference on advanced information networking and applications, pp 907–914
23. Zhu M, Liu F, Cai Z, Xu M (2015) Maintaining connectivity of MANETs through multiple unmanned aerial vehicles. Mathematical Problems in Engineering, vol 2015, p 14
24. Heimfarth T, de Araujo JP, Giacomin JC (2014) Unmanned aerial vehicle as data mule for connecting disjoint segments of wireless sensor network with unbalanced traffic. In: IEEE 17th international symposium on object/component/service-oriented real-time distributed computing, pp 246–252
25. Marinho MAM, de Freitas EP, de Costa JPCL, de Almeida ALF, de Sousa RT (2013) Using cooperative MIMO techniques and UAV relay networks to support connectivity in sparse wireless sensor networks. In: International conference on computing, management and telecommunications (ComManTel), pp 49–54, 21–24 Jan 2013
26. Guillen-Perez A, Sanchez-Iborra R, Cano MD, Sanchez-Aarnoutse JC, Garcia-Haro J (2016) WiFi networks on drones. In: 2016 ITU kaleidoscope: ICTs for a sustainable world (ITU WT), pp 1–8
27. Asadpour M, Giustiniano D, Hummel KA, Heimlicher S (2013) Characterizing 802.11n aerial communication. Presented at the proceedings of the second ACM MobiHoc workshop on airborne networks and communications, Bangalore, India
28. Yanmaz E, Yahyanejad S, Rinner B, Hellwagner H, Bettstetter C (2018) Drone networks: communications, coordination, and sensing. Ad Hoc Netw vol 68, pp 1–15
29. Ullah H, Abu-Tair M, McClean S, Nixon P, Parr G, Luo C (2017) An unmanned aerial vehicle based wireless network for bridging communication. In: The 14th international symposium on pervasive systems, algorithms and networks, pp 179–184
30. Ullah H, Abu-Tair M, McClean S, Nixon P, Parr G, Luo C (2018) UAV-based wireless network for bridging communication using IEEE 802.11 protocols. Submitted to IEEE journal on selected areas in communications (JSAC), Special issue on airborne communication networks
31. Andre T et al (2015) Application-driven design of aerial communication networks. IEEE Commun Mag 52(5):129–137
32. George SM et al (2010) DistressNet: a wireless ad hoc and sensor network architecture for situation management in disaster response. IEEE Commun Mag 48(3):128–136
33. Shengli Fu et al (2015) Spotlight: UAVs for disaster area communication. HDIAC Spotlight
34. Wu Q et al (2018) Joint trajectory and communication design for multi-UAV enabled wireless networks. IEEE Trans Wireless Commun 17(3):2109–2121
35. Gurkan Tuna et al (2012) Design strategies of unmanned aerial vehicle-aided communication for disaster recovery. In: 9th IEEE international conference on high capacity optical networks and enabling technologies (HONET)
36. Sonia Waharte et al (2010) Supporting search and rescue operations with UAVs. In: 2010 international conference on emerging security technologies (EST)
37. Silvagni Mario et al (2016) Multipurpose UAV for search and rescue operations in mountain avalanche events. Geomatics Nat Hazards Risk 8(1):18–33
38. Swiss foundation for mine action (2016) Simulation—drones for search and rescue in emergency response simulation
39. Zhang J et al (2016) Flooding disaster oriented USV & UAV system development & demonstration. In: IEEE-Oceans
40. Kakooei M et al (2017) Fusion of satellite, aircraft, and UAV data for automatic disaster damage assessment. J Remote Sens 38(8)
41. Corrado C et al (2017) Data fusion and unmanned aerial vehicles (UAVs) for first responders. In: 2017 IEEE international symposium on technologies for homeland security (HST)

42. Jiang X, Ren P, Luo C (2016) A sensor self-aware distributed consensus filter for simultaneous localization and tracking. Cognitive Comput 8(5):828–838
43. Gong Y, Luo C, Chen Z (2012) Two-path succussive relaying with hybrid demodulate and forward. IEEE Trans Veh Technol 61(5):2044–2053
44. Raptopoulos A (2013) No Roads? There is a Drone for That. TED Conference
45. Carlsson JG et al (2017) Coordinated logistics with a truck and a drone, Inform
46. Luo C, Casaseca-de-la-Higuera P, McClean S, Parr G, Ren P (2018) Characterisation of received signal strength perturbations using allan variance. IEEE Trans Aerosp Electron Syst 54(2):873–889
47. Morgenthaler S, Braun T, Zhao Z, Staub T, Anwander M (2012) UAVNet: A mobile wireless mesh network using unmanned aerial vehicles. In: Proc IEEE GLOBECOM Workshop-WiUAV, pp 1603–1608
48. Luo C, Wang Q, Wang X, Grecos C, Yang R, Ren P (2013) Exploiting selection diversity and recovering spectrum loss in wireless sensor networks with directional antennas. Globecom, 2013 IEEE, Atlanta, Georgia, 9–13 Dec 2013
49. Luo C, Nightingale J, Asemota E, Grecos C (2015) A UAV-cloud system for disaster sensing applications. In: IEEE 81st vehicular technology conference (VTC Spring), Glasgow, 11–14 May 2015
50. Luo C, McClean SI, Parr G, Teacy L, De Nardi R (2013) UAV position estimation and collision avoidance using the extended kalman filter. IEEE Trans Veh Technol 62(6):2749–2762

Human Detection Based on Radar Sensor Network in Natural Disaster

Wei Wang

Abstract In recent years, natural disasters, such as earthquakes, landslides and others, have caused significant damage to people's lives and property. Victims are often trapped in collapsed buildings. Thus the development and understanding of modern techniques for disaster relief are of immense current interest and need. As a significant advancement in wireless communication, the emerging UWB Radar Technology is a key technology that UWB is applied in object identification, which is characterized by high resolution, good anti-interference ability and strong penetrability and so on, has been widely used in various fields, including natural disaster detection, through-wall radar imaging, ground penetrating radar technology, medical imaging, target ranging and personnel positioning, disaster relief and so on. In this chapter, the author will describe some algorithms for human detection based on UWB radar sensor network in natural disaster. Firstly, we study the fuzzy pattern recognition and genetic algorithm which is used to identify the multi-status human being after the brick wall. The main characteristic parameters are selected and extracted from the received signal, and each feature parameters corresponding to a sub membership function. Through the genetic algorithm to optimize the sub membership function for constructing the membership function set. According to fuzzy pattern recognition principle of maximum degree of membership function to establish target prediction function, and used MATLAB to carry on the simulation for it. Secondly, we study the stacked denoising autoencoder algorithm in deep learning to study the through wall human target recognition under imbalanced samples of single sensor and multi-sensor data respectively. The experimental results show that the stacked denoising autoencoder algorithm in deep learning adopted herein allows more effective classification and identification of through wall human targets under imbalanced sample conditions than other algorithms, and that the identification effect with multiple sensors under a certain imbalance rate is better than that with a single sensor.

W. Wang (✉)
Tianjin Key Laboratory of Wireless Mobile Communications and Power Transmission, Tianjin Normal University, Tianjin, People's Republic of China
e-mail: weiwang@tjnu.edu.cn

T. S. Durrani et al. (eds.), *Geological Disaster Monitoring Based on Sensor Networks*, Springer Natural Hazards, https://doi.org/10.1007/978-981-13-0992-2_8

Keywords UWB radar · Through-wall human being detection
Fuzzy pattern recognition · Genetic algorithm · Imbalanced samples
Autoencoder · Deep learning

1 Background of Human Detection Based on Radar Sensor Networks

The first generation of radar was started in the 1920s at the beginning of 1924–1938 years. It used only the reflection of the electromagnetic wave to achieve some simple functions, the frequency range in the tens of megahertz, resolution and accuracy is not high, ranging in scope within one hundred km. The second generation of radar is produced in 1936–1960 years, the development of radar devices and technology is more advanced. The generation of the third generation of radar is in 1971–1990 years or so. This generation of radar uses computers, microprocessors and large scale integrated circuits to improve radar performance, reduce radar volume and weight, and improve the reliability of information processing. The fourth generation of radar came into being after 2000, and this generation of radar is to further reduce the volume and weight of the radar by using smaller, more reliable devices. The UWB technology, originally developed by the Defense Advanced Research and Planning Agency, was called baseband, non-carrier, Pulse communications, or time domain signal transmission until 1989, when the Department of Defense named it Ultra-Broadband. From the emergence of the concept to the 1990s, ultra wide band technology was primarily based on the initial impulse radio technology using pulse width of nanosecond or sub nanosecond as information carrier, and was used primarily for military radar and wireless systems with low cargo and detection rates. Ultra wideband radar can be divided into military and civilian radar in accordance with the purposes, including military radar, warning radar, radar, IFF, mainly for the detection of landmines, detecting the hidden military targets with foliage camouflage dangerous objects; and civilian radar including navigation radar, weather radar, radar, weather radar at weather disasters, especially in monitoring and early warning of sudden disaster weather plays a very important role in [1]. The advantages of UWB are high resolution and multipath fading. Large system capacity, high transmission rate; Low interference, good confidentiality; Precise positioning capability; Low power consumption, long continuous use; Small size, low cost, flexible [2].

UWB radar signal is the radar antenna electromagnetic waves emitted by any direction emission, if there are objects in the direction of the above goals, the reflection of electromagnetic wave reflection, this contains all kinds of information received by the target and the radar antenna, the receiving device to process the signal, obtaining effective information of targets. From the process of radar signal sending and receiving, it can be found that UWB signals will be received after different targets, and there are target characteristics information in the received signals, but there are obvious differences between them. Therefore, extracting the characteristic parameters required for the reaction of the target information in the received signal, which is used in some classification algorithms, classification and identification of target for the ultra wideband radar used to detect around the target, play an important role in hazardous environments or under special circumstances or buried under the wall human rescue.

UWB radar has the advantages of strong penetration, high range resolution, strong clutter suppression, high location accuracy, insensitivity to channel fading and low power consumption. It can penetrate, wall and other media, and detect, track and locate targets behind the wall. At present, the ultra wideband radar in penetration of various materials of wall on the wall after the human target detection has been widely used, has accumulated abundant research achievements in various fields, such as airport security, reconnaissance through terrorists in daily counter-terrorism operations, the police can determine the specific location of the terrorists and hostages with it to provide strong support for the arrest of terrorists and rescue the hostages, trapped in the rubble, fires, earthquakes, avalanches and other emergency situations, rescue personnel can use it to find the location of trapped persons, in order to maximize the life saving time also includes the location, medical monitoring, UWB radar technology has the application of [3–6] is widely used in military and civil fields. UWB radar is becoming more and more mature for human body target recognition.

2 State of the Art in Target Detection Based on UWB Radar Sensor Networks

Due to its own high-range resolution, wide bandwidth and other outstanding features of UWB radar technology, it has become the best choice for through wall human target recognition. In addition, the UWB pulse itself has a very high propagation energy in the frequency range of the bandwidth and strong wall penetration [7]. Through wall human target recognition has important application value in the field of military and civilian applications. The problem of through wall human target recognition has gradually become the hot spot of scholars at home and abroad. Currently, researchers have made a great deal of research achievements in the field of human target recognition.

In [8], it studied the technology of wall detection for different types of walls. In this chapter, a single-pulse UWB radar is used to detect stationary human targets. A new method based on short-time Fourier transform is proposed. The existing The clutter suppression technique based on singular value decomposition is applied to different types of walls for wall detection. In [7], it applied UWB radar to human body through the wall and proposed a method based on Fast Fourier Transform and S-Transform to detect and identify the life characteristics of human body. The center frequency of vital signal was extracted from the experimental data Accurately locate the location of human targets. Compared with other human detection through the wall, the chapter focuses on the processing and identification of life signals under strong clutter. One can use this method to search for and locate survivors trapped under buildings during earthquakes, explosions or fires. In [9], it used wall-penetrating radar to detect and locate human targets hidden behind obstacles. This chapter introduces a new adaptive detection technology for human detection based on exponential averaging with weighted coefficients. The experiment shows a moving measurement of a human body through the wall to detect and locate the target. The detection of human signal is very challenging, because the signal may be buried by noise using ultra-wideband radar sensor to study the measurement of signal through the wall, the article uses ultra-wideband radar PulsOn 220 through the human body Probing experiments, conducted through two different types of walls, gypsum walls and wooden doors, show the detection of human targets hidden behind walls or in buildings, which is of interest to rescue, surveillance and safety operations [10]. In [11, 12], it used UWB random noise radar for wall penetration monitoring and wall penetration imaging. The experimental results show that the UWB random noise radar can be used to detect and track the human being sheltered by the building wall, and at the same time can conceal the target image. In [13], it proposed an UWB radar based on the empirical mode decomposition of cross-correlation filters, which can effectively detect and locate coal miners. In [14], it developed a UWB-based UWB synthetic aperture radar with the purpose of verifying the penetration ability of the UWB radar cluster and the high-resolution imaging ability of the hidden target. In [15], it used Doppler radar systems to detect stationary human targets obstructed by walls and obstacles, and identifies the presence of human beings by detecting respiratory-induced Doppler signals and the movements of the human arm and wrist. Experimental results show that The Doppler radar system extracts the validity of the Doppler signal corresponding to human activity. In [16], it discussed UWB radar penetration detection technology. Due to the large amount of UWB radar data, compression sensing theory is introduced to collect UWB data. In this chapter, the singular values of compressed radar data are obtained by singular value decomposition and the compressed UWB radar data of the two target states of human body after gypsum wall are collected. The experimental results show that the singular value increases when there is a human target behind the wall when compared to the state without human target behind the wall. In [17], it studied ultra-wideband radar for detecting and locating people or any non-metallic obstacles that are obscured by walls and obstacles. Ultra-wideband sensor networks can be used in targeting and imaging of targets, intrusion detection of surrounding targets, in-vehicle sensing, monitoring of outdoor sports, testing of freeways and bridges, and other civil facilities [18]. Estimation algorithms based on TOA (Time of Arrival) and received signal strength are also used in UWB sensing networks to identify NLOS channels [19]. According to the statistical characteristics of multipath channels in [19], parameters such as average additional delay, mean square delay and value are extracted in the impulse response, and the joint probability density function is used to detect whether the channel is a line-of-sight channel or a non-line-of-sight channel.

3 Through-Wall Human Being Detection Based on Fuzzy Pattern Recognition

3.1 Theory

In this chapter, the UWB radar equipment is used to identify the multi-status human being after the brick wall by the fuzzy pattern recognition and genetic algorithm. Firstly, The main characteristic parameters are selected and extracted from the received signal, and each feature parameters corresponding to a sub membership function. Then, Through the genetic algorithm to optimize the sub membership function for constructing the membership function set. Lastly, According to fuzzy pattern recognition principle of maximum degree of membership function to establish target prediction function, then used MATLAB to carry on the simulation for the experiment results. The algorithm is shown in follow:

Algorithm 1: Through-wall human being detection based on fuzzy pattern recognition
Step 1: Extracting feature parameters to construct sub membership function
Step 2: Using genetic algorithm to optimize the membership function for constructing membership function set
Step 3: Construction of target prediction function by fuzzy pattern recognition algorithm
Step 4: Based on the principle of maximum membership degree realize the multi-status target recognition behind the wall

3.1.1 Selection and Extraction of Characteristic Parameters

The task of feature selection and feature extraction is to determine the data which are meaningful to the classification as characteristic data according to the measured data. These data can not only reflect the similarity of the same pattern, but also reflect the difference of different types of models. The purpose is to constitute characteristics for recognition and identification through feature selection and extraction and to preserve the classified information as much as possible while ensuring certain classification accuracy. The characteristics of recognition objects include physical features, structural features and mathematical features. In this chapter, the mathematical features of data are extracted as the characteristic parameters of through wall multi-status human being detection. There are six kinds of status in through-wall recognition, each status has its corresponding characteristic parameters which mainly comprise the kurtosis, skewness, energy, maximum amplitude, variance and covariance of the received signal. Kurtosis is a measure of the "tailedness" of the probability distribution of a real-valued random variable as shown in formula (3.1). Skewness is a measure of the asymmetry of the probability distribution of a real-valued random variable about its mean as shown in formula

(3.2). This chapter selected a total of 36 characteristic parameters, the characteristic parameters of each status are defined as follows:

$$K = \frac{1}{\sigma_{|r|}^4 T} \int \left[\left| r(t) - \mu_{|r|} \right| \right]^4 dt \tag{3.1}$$

$$S = \frac{\frac{1}{T} \int \left[\left| r(t) - \mu_{|r|} \right| \right]^3 dt}{\left[\sqrt{\frac{1}{T} \int \left[\left| r(t) - \mu_{|r|} \right| \right]^2 dt} \right]^3} \tag{3.2}$$

$$G_r = \int_{-\infty}^{+\infty} |r(t)|^2 dt \tag{3.3}$$

$$r_{max} = \max_t |r(t)| \tag{3.4}$$

$$\sigma_{|r|}^2 = \frac{1}{T} \int \left[\left| r(t) - \mu_{|r|} \right| \right]^2 dt \tag{3.5}$$

$$cov = \frac{1}{T} \int \left[\left| \left(r_1(t) - \mu_{1_{|r|}} \right) \left(r_2(t) - \mu_{2_{|r|}} \right) \right| \right] dt \tag{3.6}$$

By extracting the characteristic parameters of the received signal, a set of membership functions for target recognition is constructed.

3.1.2 Genetic Algorithm Theory

The basic idea of genetic algorithm is to encode the optimized parameters first and then process the individuals obtained after the encoding. Therefore, the genetic algorithm can not only optimize and solve the traditional objective function, but also can optimize and solve the structural object, such as using the genetic algorithm to optimize the matrix and graphics [20]. The fitness function is used to evaluate multiple parameters of the search space at the same time, each possible problem is expressed as a chromosome, and then the selection, crossover and mutation operations are performed according to the genetic laws until the termination condition is satisfied.

Since the fuzzy concepts in practical application are various, we cannot define a generic membership function to represent all cases in practical applications. So the quality of membership selection does not have a unified evaluation standard at present. The membership functions commonly used in MATLAB include triangular membership function, trapezoidal membership function, Gaussian membership function, B-Gaussian membership function, bell-shaped membership function, etc. After comprehensive consideration and practical analysis, this chapter selects the Gaussian function as the sub-membership function. The Gaussian function is defined as follows:

$$f(x) = e^{\frac{-(x-a)^2}{b^2}} \tag{3.7}$$

Wherein a is the mean, b is the variance, both parameters can be obtained by the genetic algorithm. The membership function set includes 36 sub-membership functions, and 72 parameters need to be optimized. The formula of the membership function set is as follows:

$$F = \begin{bmatrix} e^{\frac{-(x-a_{11})^2}{b_{11}^2}} & e^{\frac{-(x-a_{12})^2}{b_{12}^2}} & e^{\frac{-(x-a_{13})^2}{b_{13}^2}} & e^{\frac{-(x-a_{14})^2}{b_{14}^2}} & e^{\frac{-(x-a_{15})^2}{b_{15}^2}} & e^{\frac{-(x-a_{16})^2}{b_{16}^2}} \\ e^{\frac{-(x-a_{21})^2}{b_{21}^2}} & e^{\frac{-(x-a_{22})^2}{b_{22}^2}} & e^{\frac{-(x-a_{23})^2}{b_{23}^2}} & e^{\frac{-(x-a_{24})^2}{b_{24}^2}} & e^{\frac{-(x-a_{25})^2}{b_{25}^2}} & e^{\frac{-(x-a_{26})^2}{b_{26}^2}} \\ e^{\frac{-(x-a_{31})^2}{b_{31}^2}} & e^{\frac{-(x-a_{32})^2}{b_{32}^2}} & e^{\frac{-(x-a_{33})^2}{b_{33}^2}} & e^{\frac{-(x-a_{34})^2}{b_{34}^2}} & e^{\frac{-(x-a_{35})^2}{b_{35}^2}} & e^{\frac{-(x-a_{36})^2}{b_{36}^2}} \\ e^{\frac{-(x-a_{41})^2}{b_{41}^2}} & e^{\frac{-(x-a_{42})^2}{b_{42}^2}} & e^{\frac{-(x-a_{43})^2}{b_{43}^2}} & e^{\frac{-(x-a_{44})^2}{b_{44}^2}} & e^{\frac{-(x-a_{45})^2}{b_{45}^2}} & e^{\frac{-(x-a_{46})^2}{b_{46}^2}} \\ e^{\frac{-(x-a_{51})^2}{b_{51}^2}} & e^{\frac{-(x-a_{52})^2}{b_{52}^2}} & e^{\frac{-(x-a_{53})^2}{b_{53}^2}} & e^{\frac{-(x-a_{54})^2}{b_{54}^2}} & e^{\frac{-(x-a_{55})^2}{b_{55}^2}} & e^{\frac{-(x-a_{56})^2}{b_{56}^2}} \\ e^{\frac{-(x-a_{61})^2}{b_{61}^2}} & e^{\frac{-(x-a_{62})^2}{b_{62}^2}} & e^{\frac{-(x-a_{63})^2}{b_{63}^2}} & e^{\frac{-(x-a_{64})^2}{b_{64}^2}} & e^{\frac{-(x-a_{65})^2}{b_{65}^2}} & e^{\frac{-(x-a_{66})^2}{b_{66}^2}} \end{bmatrix} \tag{3.8}$$

The mean and variance of the Gaussian function are optimized by the genetic algorithm. In this chapter, the population type of the genetic algorithm is double vectors. Suppose the number of the initial population is 150, the variables that need to be optimized are the mean and variance, which represent genes in the heritage algorithm, 72 in total. The range of each variable is 0.1–1.5. The fitness function is defined as follows:

$$\text{Fitness Function} = \frac{1}{\sum_{i=1}^{n}\left(\frac{right\_w_i}{total\_w_i}\right)+\sum_{i=1}^{n}\left(\frac{right\_h_i}{total\_h_i}\right)+\sum_{i=1}^{n}\left(\frac{right\_j_i}{total\_j_i}\right)+\sum_{i=1}^{n}\left(\frac{right\_z_i}{total\_z_i}\right)+\sum_{i=1}^{n}\left(\frac{right\_k_i}{total\_k_i}\right)+\sum_{i=1}^{n}\left(\frac{right\_s_i}{total\_s_i}\right)} \tag{3.9}$$

Wherein $right\_w_i$ refers to the *i*th correctly recognized through-wall no person status. Similarly, $right\_h_i$ refers to the through-wall normal breathing status, $right\_j_i$ refers to the through-wall two person normal breathing status, $right\_z_i$ refers to the through-wall two persons walking 2 m away status, $right\_k_i$ refers to the through-wall swing arms status, and $right\_s_i$ refers to the through-wall three persons normal breathing status. Since this chapter uses MATLAB to realize the genetic algorithm, the fitness function is optimized according to the minimum value, therefore this chapter chooses the reciprocal of the total recognition accuracy of all status as fitness function. Individual generic selection is conducted according to the fitness function. In the genetic algorithm, selection, crossover and mutation are the most basic operations. We choose the roulette algorithm as the selection algorithm and select the single point crossover algorithm, the crossover probability being 0.8, the mutation probability being 0.2, and the genetic algebra being 100 generations. The termination condition is that the genetic algorithm will terminate when it reaches the genetic algebra or the fitness value of the parameter individual reaches the ideal result.

3.1.3 Fuzzy Pattern Recognition Theory

The basis of fuzzy pattern recognition is fuzzy mathematics which has only 40 years of history since its inception in 1965. Fuzzy pattern recognition has been an active research field of fuzzy application since the birth of fuzzy mathematics. The research contents comprise computer image recognition, handwritten text automatic recognition, cancer cell recognition, white blood cell identification and classification, disease prediction, classification of various types of information and so on [21].

Definition of fuzzy set: a fuzzy subset A in a given domain of discourse refers that for any $x \in X$, a number $\mu_A(x)$ is determined. $\mu_A(x)$ is referred to as the degree of membership of x for the fuzzy set A, and $\mu_A(x) \in [0,1]$. Mapping $\mu_A(x) : X \to [0,1], x \to \mu_A(x)$. $\mu_A(x)$ is referred to as the membership function of A, which is used to describe the degree of membership of the factors in the subset A for A. The value of the membership function is referred to as the degree of membership. The larger the degree of membership is, the higher the degree that x affiliates to A will be. The fuzzy subset is usually referred to as the fuzzy set or fuzzy sets.

Representation method of the fuzzy set: In the practical application, there are many ways to represent the fuzzy set. In principle, it is required to show the relationship between all the elements in the domain and its corresponding membership degree. Such methods include the summation representation, the integral representation, ordered pair representation, vector notation and other methods. Methods of Fuzzy Pattern Classification: (1) Direct Method, The method of directly judging the affiliation of the sample by calculating its membership degree is referred to as the membership principle of the pattern classification. Supposing that there are N fuzzy sets $A_1, A_2, \ldots, A_N$ in the domain X and each fuzzy set A_i has a membership function $\mu_{A_i}(x)$, then for any $x \in X$, if

$$\mu_{A_i}(x) = \max\left[\mu_{A_1}(x), \mu_{A_2}(x), \ldots, \mu_{A_N}(x)\right] \tag{3.10}$$

It will be considered that x is subordinate to A_i. The membership principle is also referred to as the direct method of fuzzy pattern recognition, which is used for recognition of a single pattern. (2) Indirect Method, unlike the membership principle, the near-selection principle is a method for group recognition. Supposing that there are n known types of fuzzy subsets $A_1, A_2, \ldots, A_n$ in the domain x, if $i \in \{1, 2, \ldots, n\}$, then

$$\sigma(B, A_i) = \mathrm{Max}\,\sigma(B, A_j) \tag{3.11}$$

Wherein $\sigma(A, B) = 1 - C[d(A, B)]^d$, C and d are two properly selected parameters, $d(A, B)$ can be different distances. Relative to $A_1, A_2, \ldots, A_{i-1}, A_{i+1}, \ldots, A_u$, if B is closest to A_i, B belongs to the category of A_i pattern.

In this chapter, the method of fuzzy pattern classification is a direct method. The characteristic parameters are extracted from the received signal, and six feature parameters are extracted from each status, including kurtosis, skewness, the maximum amplitude of the received signal, the variance of the received signal and the covariance. The Gaussian function is selected as the sub-membership function, each characteristic parameter of each scene corresponds to a sub-membership function $f(x)$, thus six scenes have a total of 36 sub-membership functions, as shown in Table 1.

In this chapter, six kinds of scenes are set up, and six characteristic parameters are extracted in each scene. Therefore, the target is represented by six characteristic parameters, namely $S_c = [S_{c_1}, S_{c_2}, S_{c_3}, S_{c_4}, S_{c_5}, S_{c_6}]$, $c = 1, 2, \ldots, 6$, where S_{c_1} represents Kurtosis, S_{c_2} represents Skewness, S_{c_3} represents the maximum amplitude of the received signal, S_{c_4} represents the energy of the received signal, S_{c_5} represents the variance of the received signal, and S_{c_6} represents the covariance of the received signal. According to the mean value a and the variance b of the sub-membership functions obtained by the genetic algorithm, the membership function set is constructed by the sub-membership functions. The membership function set F is as follows:

Table 1 Corresponding relations of sub-membership functions

	No person status	Normal breathing status	Swing arms status	Two person normal breathing status	Two person walking 2 m away status	Three person normal breathing status
Kurtosis	$f_{11}(x)$	$f_{12}(x)$	$f_{13}(x)$	$f_{14}(x)$	$f_{15}(x)$	$f_{16}(x)$
Skewness	$f_{21}(x)$	$f_{22}(x)$	$f_{23}(x)$	$f_{24}(x)$	$f_{25}(x)$	$f_{26}(x)$
Amplitude	$f_{31}(x)$	$f_{32}(x)$	$f_{33}(x)$	$f_{34}(x)$	$f_{35}(x)$	$f_{36}(x)$
Energy	$f_{41}(x)$	$f_{42}(x)$	$f_{43}(x)$	$f_{44}(x)$	$f_{45}(x)$	$f_{46}(x)$
Variance	$f_{51}(x)$	$f_{52}(x)$	$f_{53}(x)$	$f_{54}(x)$	$f_{55}(x)$	$f_{56}(x)$
Covariance	$f_{61}(x)$	$f_{62}(x)$	$f_{63}(x)$	$f_{64}(x)$	$f_{65}(x)$	$f_{66}(x)$

$$F = \begin{bmatrix} f_{11}(x) & f_{12}(x) & f_{13}(x) & f_{14}(x) & f_{15}(x) & f_{16}(x) \\ f_{21}(x) & f_{22}(x) & f_{23}(x) & f_{24}(x) & f_{25}(x) & f_{26}(x) \\ f_{31}(x) & f_{32}(x) & f_{33}(x) & f_{34}(x) & f_{35}(x) & f_{36}(x) \\ f_{41}(x) & f_{42}(x) & f_{43}(x) & f_{44}(x) & f_{45}(x) & f_{46}(x) \\ f_{51}(x) & f_{52}(x) & f_{53}(x) & f_{54}(x) & f_{55}(x) & f_{56}(x) \\ f_{61}(x) & f_{62}(x) & f_{63}(x) & f_{64}(x) & f_{65}(x) & f_{66}(x) \end{bmatrix} \tag{3.12}$$

Substitute F into the target prediction formula to obtain the prediction function Y:

$$Y = S_c * F = \begin{bmatrix} S_{c_1} \\ S_{c_2} \\ S_{c_3} \\ S_{c_4} \\ S_{c_5} \\ S_{c_6} \end{bmatrix}^T * \begin{bmatrix} f_{11}(x) & f_{12}(x) & f_{13}(x) & f_{14}(x) & f_{15}(x) & f_{16}(x) \\ f_{21}(x) & f_{22}(x) & f_{23}(x) & f_{24}(x) & f_{25}(x) & f_{26}(x) \\ f_{31}(x) & f_{32}(x) & f_{33}(x) & f_{34}(x) & f_{35}(x) & f_{36}(x) \\ f_{41}(x) & f_{42}(x) & f_{43}(x) & f_{44}(x) & f_{45}(x) & f_{46}(x) \\ f_{51}(x) & f_{52}(x) & f_{53}(x) & f_{54}(x) & f_{55}(x) & f_{56}(x) \\ f_{61}(x) & f_{62}(x) & f_{63}(x) & f_{64}(x) & f_{65}(x) & f_{66}(x) \end{bmatrix} \tag{3.13}$$

Y is a matrix in 1 * 6. Elements in Y are $[Y_1, Y_2, Y_3, Y_4, Y_5, Y_6]$. According to the maximum membership principle, if

$$y = \max(Y_i) \tag{3.14}$$

It is determined that the data y is subordinate to Y_i. In this chapter, UWB radar mainly extracts the characteristic parameters related to the target from the received signal as the input signal of target recognition in the first place. The characteristic parameters extracted in this chapter mainly comprise kurtosis, skewness, energy, maximum amplitude value, variance and covariance. Then, the Gaussian function is chosen as the membership function, the mean and variance of the Gaussian function are optimized by the genetic algorithm, the membership function set of the target recognition is constructed by the characteristic parameters. Finally the target recognition category is determined according to the maximum membership degree of the fuzzy pattern recognition theory to complete the through-wall multi-status human target recognition.

3.2 Experiments and Analysis

3.2.1 Experimental System

In this chapter, six kinds of scenes are set up, including the through-wall no person status, normal breathing status of one person, swing arms status of one person, normal breathing status of two persons, waking 2 m away status of two persons and

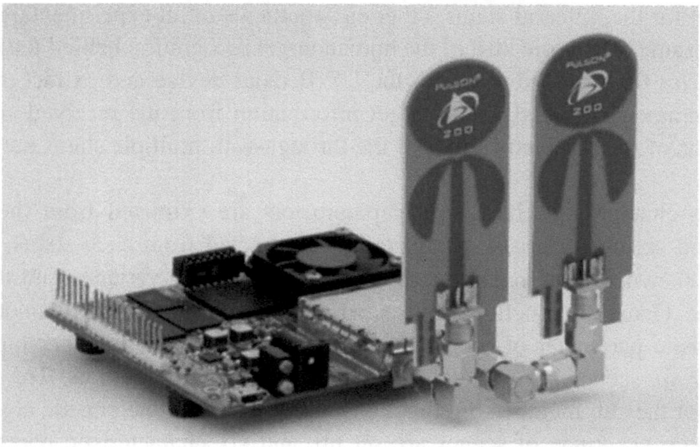

Fig. 1 Time domain

normal breathing status of three persons. Experiments mainly adopt the fuzzy pattern recognition algorithm and the genetic algorithm to identify multiple status of human being behind the wall. The experimental device used in this chapter is the P410 MRM radar device from Time Domain, as shown in Fig. 1. The device's frequency range is 3.1–5.3 GHz, the center frequency is 4.3 GHz, and PC adopts the Windows7 system. In all of the above scenes, the human targets are 1 m away from the wall, and the P410 radar device is 20 cm away from the wall. The radar device is placed on a tripod, and its distance from the ground is half of height of the measured wall, as shown in Fig. 2.

Fig. 2 UWB P410 in monostatic mode

Except for the different status, other environments of all experimental scenes are set to the same. The basic idea of the human target recognition behind the brick wall is to process the received signal of the UWB radar device and extract the characteristic parameters related to the target information from the received signal. The waveforms of the received signals of the through-wall multiple status are as shown in Fig. 3.

In this chapter, the characteristic parameters are extracted from the received signal, and six characteristic parameters are extracted from each status, including kurtosis, skewness, the maximum amplitude, energy, the variance and the covariance. The Gaussian function is selected as the sub-membership function, each characteristic parameter of each scene corresponds to a sub-membership function $f(x)$, thus six scenes have a total of 36 sub-membership functions. The mean and variance of the sub-membership functions are obtained by the genetic algorithm. In this chapter, six kinds of scenes are set up, and six characteristic parameters are extracted in each scene. Therefore, the target is represented by six characteristic parameters, namely $S_c = [S_{c_1}, S_{c_2}, S_{c_3}, S_{c_4}, S_{c_5}, S_{c_6}], c = 1, 2, \ldots, 6$, where S_{c_1} represents Kurtosis, S_{c_2} represents Skewness, S_{c_3} represents the maximum amplitude of the received signal, S_{c_4} represents the energy of the received signal, S_{c_5} represents the variance of the received signal, and S_{c_6} represents the covariance of the received signal. The optimized mean value a and variance b of the sub-membership functions obtained by the genetic algorithm, the distribution is as shown in Fig. 4. Respectively substituting a and b into formula (3.13) to get the membership function set F, then we substituting the measured data S into formula (3.10) and determining the through-wall status of the measured S according to the maximum membership principle of formula (3.11).

3.2.2 Results and Analysis

In this chapter, six different status of through wall human being are selected for data analysis. Under each scene, one set of data is randomly selected as shown in Table 2, in which S_1 is the through-wall no person status, S_2 is the through-wall normal breathing status, S_3 is the through-wall swing arms status, S_4 is the through wall normal breathing status of two persons, S_5 is the through-wall walking 2 m away status of two persons, S_6 is the through-wall normal breathing status of three persons. The data of the above five types of status are substituted into the target prediction function, the results of the identification as shown in Table 3.

We known the type of data of six groups with S_1, S_2, S_3, S_4, S_5 and S_6, then we are substituted the six groups into the target prediction function of this chapter. The data are analyzed according to the maximum membership principle of the fuzzy pattern recognition. Wherein among the six data of Y_1, 1.4338 is the largest, and it is judged as the through-wall no person status and the recognition result is correct; among Y_2, 2.0204 is the largest, it is judged as through-wall normal breathing status and the recognition result is correct; among Y_3, 2.7567 is the largest, it is

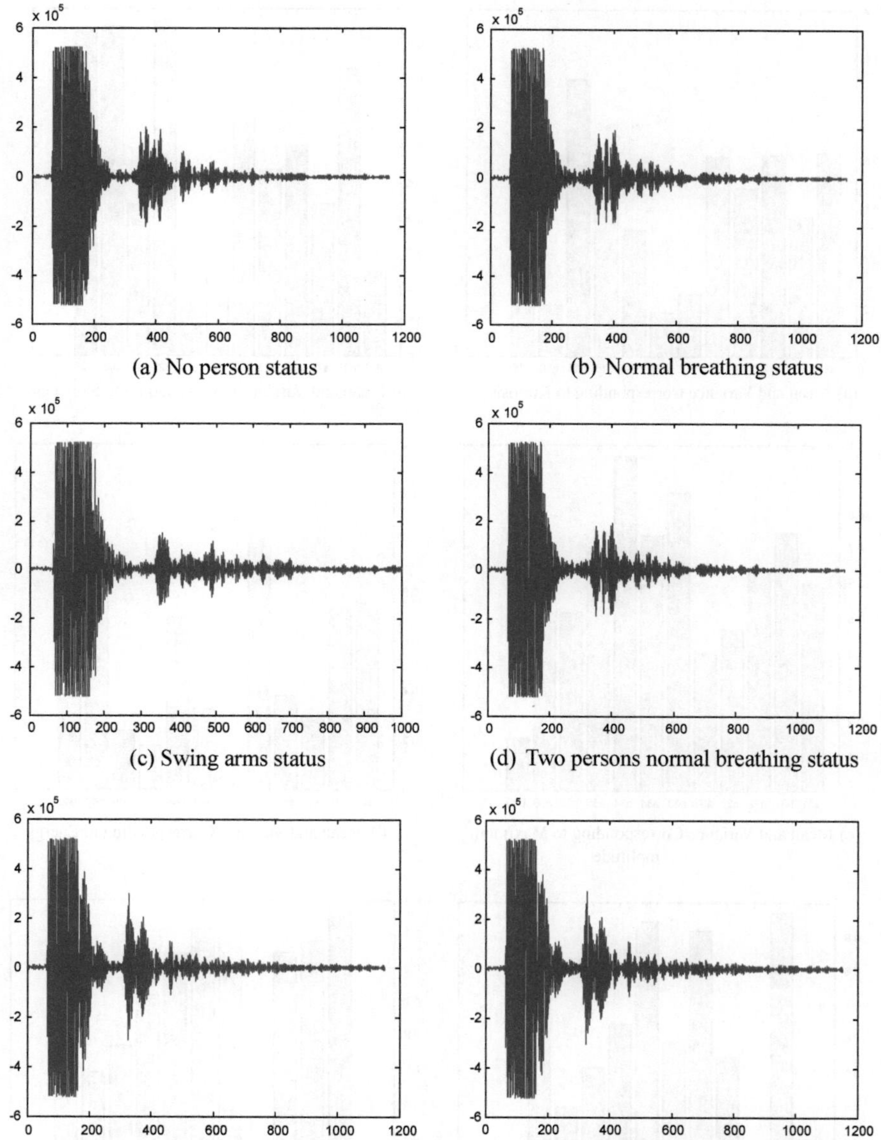

(a) No person status

(b) Normal breathing status

(c) Swing arms status

(d) Two persons normal breathing status

(e) Two persons walking 2m away status

(f) Three persons normal breathing status

Fig. 3 The waveforms of the received signal

judged as the swing arms status, and the recognition result is correct; among Y_4, 1.9121 is the largest, it is determined as the normal breathing status of two persons and the recognition result is correct; among Y_5, 0.3845 is the largest, it is determined as the status of walking around of two persons and the recognition result is correct; among Y_6, 2.5262 is the largest, it is determined as the swing arms status

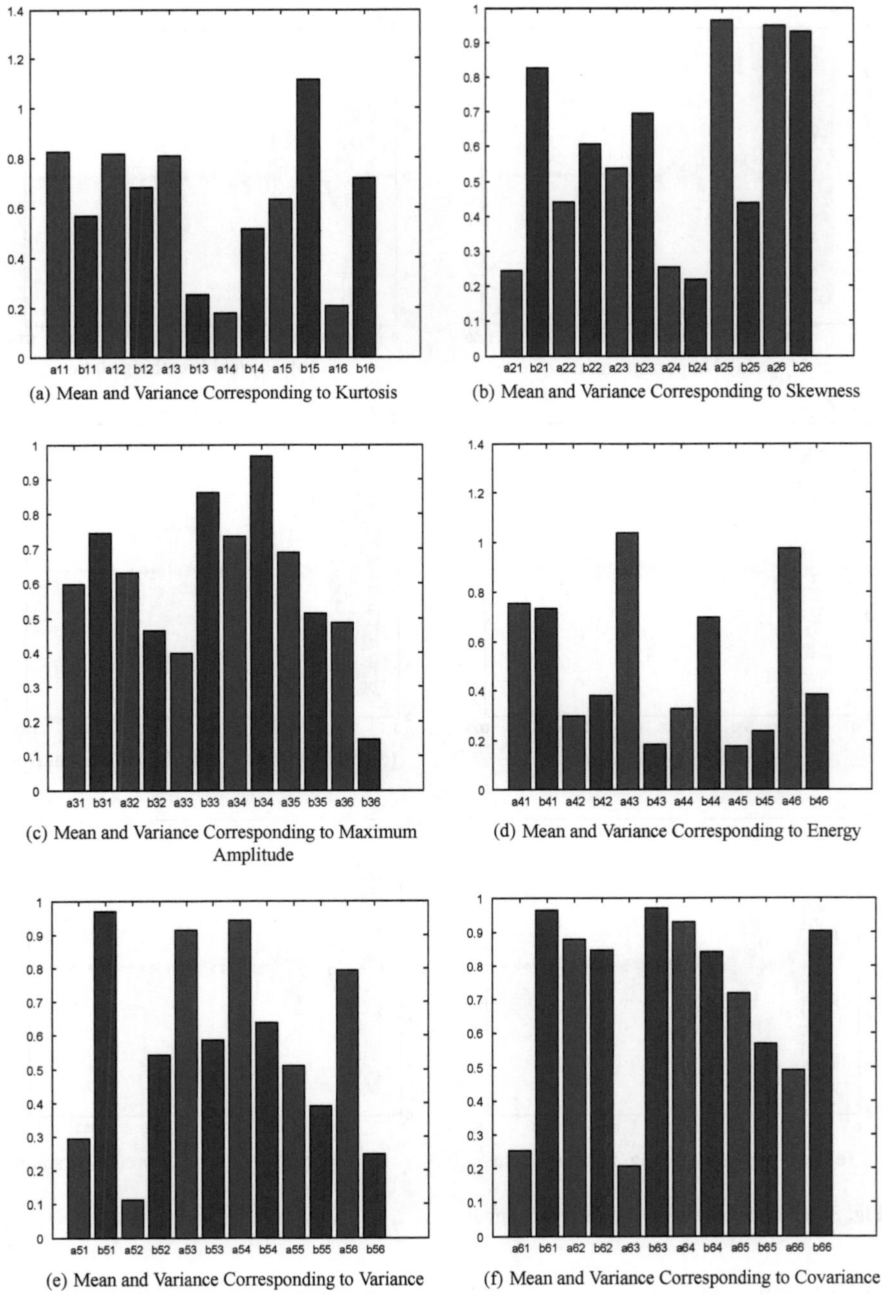

(a) Mean and Variance Corresponding to Kurtosis

(b) Mean and Variance Corresponding to Skewness

(c) Mean and Variance Corresponding to Maximum Amplitude

(d) Mean and Variance Corresponding to Energy

(e) Mean and Variance Corresponding to Variance

(f) Mean and Variance Corresponding to Covariance

Fig. 4 Mean and variance distribution diagram of sub-membership function

Table 2 Randomly selected data from a known status

Status	Kurtosis	Skewness	Maximum amplitude	Energy	Variance	Covariance
S_1	0.3916	0.0545	0.1262	0.8923	0.0029	0.0535
S_2	0.4518	−0.0430	0.4630	0.5444	0.0011	0.7383
S_3	0.8922	0.1159	0.8988	0.6742	0.4756	0.0383
S_4	0.2274	−0.0504	0.4796	0.3951	0.4857	0.6910
S_5	0.0696	0.2828	0.0229	0	0.0241	0.0470
S_6	0.3054	−0.0029	0.5150	0.9214	0.5	0.7178

Table 3 The results of the identification

Status	No person	Normal breathing	Swing arms	Two person normal breathing	Two person walking 2 m away	Three person normal breathing
Y_1	1.4338	1.2487	1.3440	0.9326	1.1730	1.1258
Y_2	1.9824	2.0204	1.7737	1.6354	1.6437	1.3169
Y_3	2.4376	2.3375	2.7567	2.0814	2.2467	2.2861
Y_4	1.5929	1.6617	1.8918	1.9121	1.8174	1.5876
Y_5	0.1270	0.2477	0.2802	0.2366	0.3845	0.2558
Y_6	2.2455	2.2163	2.5262	2.3983	2.4479	2.1962

according to the maximum membership principle. However, since S_6 belongs to the normal breathing status of three persons, thus the judgment result is incorrect. Due to the fact that the difference between normal breathing of three persons and the swing arms status is small, there is the possibility of wring judgment. This chapter considers that the normal breathing status of three persons wrongly determined as the swing arms status is still in a reasonable range. In this chapter, the data of six kinds of through-wall status are randomly selected to conduct the analysis.

In this chapter, 500 sets of data are received for testing each status, and each set of data contains 1152 data points. In the 500 sets of data for each status, we uniformly choose 50 sets of data as the training data and verification data, a total of 300 sets of data, and these 300 sets of verification data are substituted into the target prediction function of the fuzzy pattern recognition algorithm proposed in this chapter. Wherein the recognition accuracy of the through-wall no person status is 98%; the recognition accuracy of the through-wall normal breathing status is 82%; the recognition accuracy of the through-wall swing arms status is 96%; the recognition accuracy of the through-wall two person normal breathing status is 94%; and the recognition accuracy of the through-wall two person walking 2 m away status is 96%. This chapter successfully and accurately recognized five different through-wall status. The ROC curve is utilized to analyze the experiment result which shows that the ROC curve is closer to the upper left corner. It represents that the algorithm has a high precision. The ROC curve of the pattern recognition conducted to multiple through-wall status through the fuzzy pattern

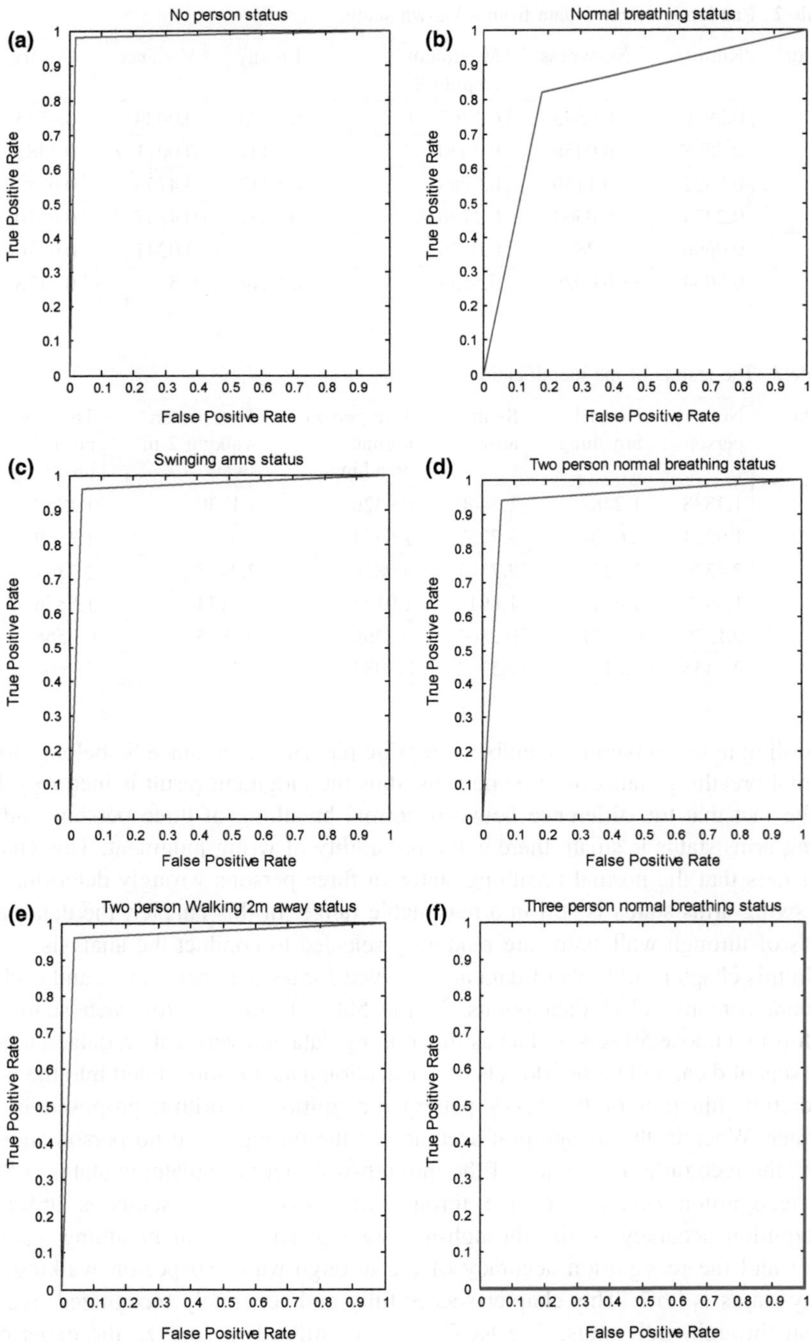

Fig. 5 ROC graphs of five through-wall status

Table 4 Comparison result of classification algorithm

AUC	Fuzzy pattern recognition	KNN	J48	ZeroR
No person status	0.98	0.4645	NaN	0.4349
Normal breathing status	0.82	0.5198	0.88	0.4538
Swing arms status	0.96	0.6648	0.8462	0.4349
Two person normal breathing status	0.94	1	1	0.4414
Two person walking 2 m away status	0.96	0.824	0.587	0.4414
Three person normal breathing status	0	NaN	NaN	0.4539

recognition algorithm is shown in Fig. 5. The classification effect of the first five status has achieved the prospective target:

In this chapter, We use classification algorithm of data mining tool to compare with the fuzzy pattern recognition algorithm. WEKA (Waikato Environment for Knowledge Analysis) is a data mining tool. There are a lot of different classification algorithms including the decision trees algorithm, Naive Bayes algorithm, Logistic algorithm, AdaBoost algorithm, artificial Neural network algorithm, KNN algorithm and so on [22]. We choose KNN, J48 and ZeroR in WEKA to compare with ours in this chapter under the condition that the data training set is completely identical to the testing set. The experimental results are shown in Table 4. It can be clearly seen from Table 4 the target recognition based on the through-wall multi-status of human being. According to the comparison of recognized status amount, the fuzzy pattern recognition algorithm proposed in this chapter can better recognize five types of through-wall human status. The KNN classification algorithm in Weka recognized five status, but the recognition accuracy of each status was far lower than that of the fuzzy pattern recognition algorithm proposed in this chapter; J48 recognized four status; ZeroR recognized six status, but compared with the first five status, the accuracy of the algorithm is much lower than that of the algorithm proposed in this chapter. According to the comparison based on the recognition accuracy, the accuracy of the algorithm proposed in this chapter is higher than the other three algorithms. Therefore, the proposed algorithm is obviously superior to the other three algorithms.

3.3 Conclusions

We mainly processes with the signals of the receiving end of the UWB radar device, and extracts the characteristic parameters related to target information, including kurtosis, skewness, energy of the received signal, amplitude value, variance and covariance. Then these six characteristic parameters are substituted into the fuzzy pattern recognition algorithm and the genetic algorithm as the target parameters to construct the target prediction function needed for recognition and recognize the target according to the maximum membership principle of fuzzy pattern recognition. Experimental results show that the algorithm can effectively distinguish the through-wall no person status, the through-wall normal breathing status, the through-wall swing arms status, the through-wall two person normal breathing status and the through-wall status of walking around of two persons. The results have very important theoretical significance and practical application value. The difference between the through-wall slow breathing status of three persons and the swing arms status is small, so there is the possibility of wrong judgment, we will do further research. This chapter compares the fuzzy pattern recognition algorithm with KNN, J48 and ZeroR, and it can be clearly seen that the proposed algorithm is superior to the other three algorithms. A more effective algorithm will be established in the next step to analyze and process various through-wall status and enables it to be applied in the complicated actual environment.

4 Through Wall Human Target Recognition Under Imbalanced Sample Based on Deep Learning

This section we proposed to apply the autoencoder algorithm in the deep learning network model to the through wall human target recognition under imbalanced sample conditions. The autoencoder algorithm extracts the concise data feature expression by automatically learning the intrinsic features in the data. Autoencoder network based on the addition of denoising coding and sparse constraints and other conditions to extract more effective feature expression to improve classification recognition rate. Then the training set data is input to the depth network for training. After the model training is completed, the test set data are input into the trained network for testing, and the classification result is obtained.

4.1 Network Construction and Training

In this section, we used a four-layer stacked denoising autoencoder network consisting of an input layer, two hidden layers and an output layer to perform experiments, as shown in Fig. 6. This section of the stacked denoising autoencoder network training process: First, by layer-by- layer greedy training method in turn training each layer of the network, the entire depth of the neural network pre-training, that is, the first use of the original input to train the first layer of autoencoder network parameters, and then the output of the hidden layer of the autoencoder as the next input of an autoencoder, so that one by one training layer by layer learning network parameters of each layer, after unsupervised pre-training, in order to allow autoencoder network has the function of classification and identification needs to add a classifier after the last hidden layer of the autoencoder neural network, and then use supervised learning to adjust the parameters of the classifier by using the labeled samples to fine-tune and minimize the error of the

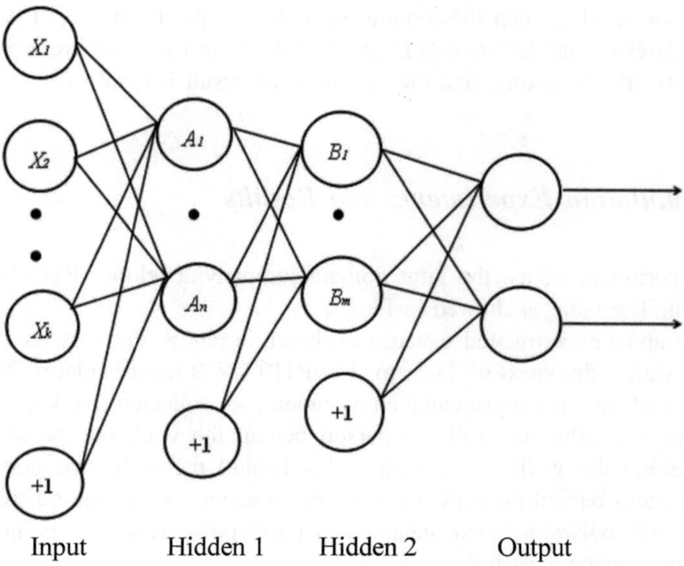

Fig. 6 Stacked denoising autoencoder network structure

prediction target. Constantly adjust the parameters of the entire network. In addition, it is also possible to fine-tune the parameters of all layers and improve the result simultaneously by supervised learning of a multi-layer neural network. Network training steps are as follows:

Algorithm 2: Stacked denoising autoencoder
Input: training set, testing set.
Output: testing error rate.
(1) construct a four-layer stacked denoising autoencoder deep network;
(2) feed the data into the first layer of the network, so that the output is equal to the input;
(3) minimize the reconstruction error between the original input and the reconstructed output to obtain the first hidden layer;
(4) output of the first hidden layer is used as the input of the next layer, and the input is equal to the output;
(5) repeat step 3, obtain the second hidden layer;
(6) output of the second hidden layer is used as the input of the classifier;
(7) use the labeled data to fine-tune the network and minimize the reconstruction error;
(8) get the test result.

In order to improve the performance of the network, the number of neuron nodes in each layer, noise parameters, sparse penalty parameters, batches and iterations are set according to the actual situation of the experiment. In addition, we also added regularization restrictions and Dropout technology to improve the classification results. Then the training set data is input to the depth network for training. After the model training is completed, the test set data are input into the trained network for testing, and the classification result is obtained.

4.2 Simulation Experiments and Results

In this experiment, we use the Time domain company developed P410 UWB radar equipment. The radar is showed in Fig. 7.

We set up an experimental scenario as shown in Fig. 8. The material of the wall is a brick wall, a thickness of 23.5 cm. The P410 UWB radar is placed 20 cm from the brick wall. In this experimental environment, we collected five kinds of human body state data after the wall: no person behind the wall; one person breathes slowly behind the wall; one person walks behind the wall; two persons rapid breathing status behind the wall; and one person waving status behind the wall. We collected 500 pulses in each state and set the pulse sampling point to 1000. Experimental scene is as follows:

In practical applications, such as detecting terrorists through walls and judging the specific positions of terrorists and hostages in daily anti-terrorist operations, search and rescue personnel are looking for trapped persons in emergencies such as fires, earthquakes and avalanches. In these applications, the data for someone is crucial to us. We selects the experimental data from the collected five kinds of state data as a training set and a test set, in which we set as the majority of categories in the unmanned state, the other categories for the minority categories, we will make the experiment unattended data quantity more than any other state, so our experimental dataset is an imbalanced dataset. The range of some data may be particularly large, resulting in slow network convergence and long training time. Before training

Fig. 7 P410 UWB radar

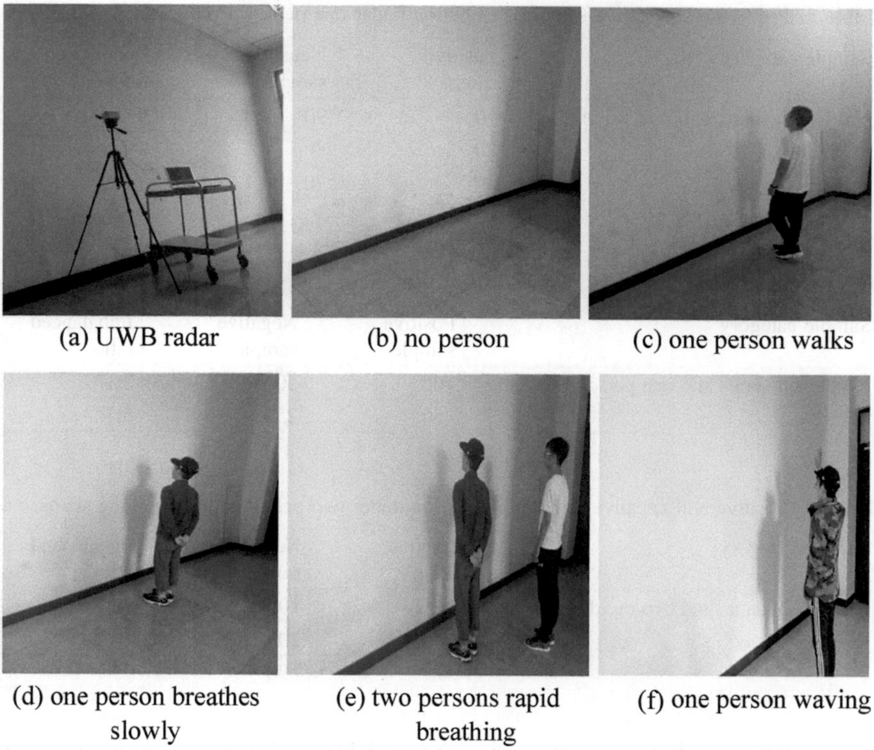

Fig. 8 Experimental scene

the network generally need to normalize the data processing. At the same time, in order to facilitate training and improve training speed, the entire data set can be batch-processed, that is, multiple sample data can be integrated into data blocks. If there are 1000 samples, each sample is 200 dimensions, 100 samples of data can be integrated into one data block, the size of each data block is 100 * 200, there are 10 such data blocks to form a three-dimensional array, Size is 100 * 200 * 10. Research shows that modularizing data can improve convergence speed.

From the collected multi-state data, five types of human body state data are selected as the training set and the test set respectively. Since we define the majority of samples when the unmanned state behind the wall is based on experiments, the number of unmanned state samples in the experiment is greater than that of several other states. In other words, the sample when no one is behind the wall is the majority, and the other states are the minority. In order to test the algorithm, we chose the most class (negative class) data for the unmanned state behind the wall, one waved behind the wall, one slow breathing, two rapid breathing state for the minority class (positive class). We used MATLAB software tools as experimental platform. The distribution of the experimental data samples for various states is shown in Tables 5, 6, and 7

Table 5 Positive and negative sample distribution under one person waving status

Sample category	Positive sample	Negative sample	Imbalanced ratio
No person status and one waving status	10	90	9:1
	20	80	4:1
	30	70	7:3

Table 6 Postitive and negative sample disitribution under one person breathing slowly status

Sample category	Positive sample	Negative sample	Imbalanced ratio
No person status and one person breathes slowly	100	200	2:1
	50	250	5:1

Table 7 Positive and negative sample distribution under two person rapid breathing status

Sample category	Positive sample	Negative sample	Imbalanced ratio
No person status and two rapid breathing status	10	190	19:1
	20	200	10:1
	30	90	3:1

In different imbalanced ratios, we choose KNN and J48 algorithm compared with the autoencoder we use, and we used the ROC curve to evaluate the experimental results. The ROC curve results of the three algorithms under different imbalance ratios are shown in Fig. 9. The closer the ROC curve is to the upper left corner, the higher the accuracy of the classification algorithm. It can be seen from Fig. 9 that under different imbalance ratios, the autoencoder algorithm has a better classification effect than other algorithms.

4.3 Multi-sensor Data Simulation and Results

In the distance of four meters in length, moving the equidistance radar, collected at 40 cm every two walls after the unmanned state and the wall when a person walking around the data, as a point of data collected a total of 10 points two states of the experimental data. In this experimental environment, we collected 500 pulses in each state and set the pulse sampling point to 1000. We use a few samples (positive) when no one is behind the wall, and a majority sample (negative) when one moves around the wall. During the experimental data selection, the multi-sensor data is distributed under a single sensor. The distribution of multi-sensor data positive and negative samples is shown in Table 8. The single-sensor data distribution is shown in Table 9.

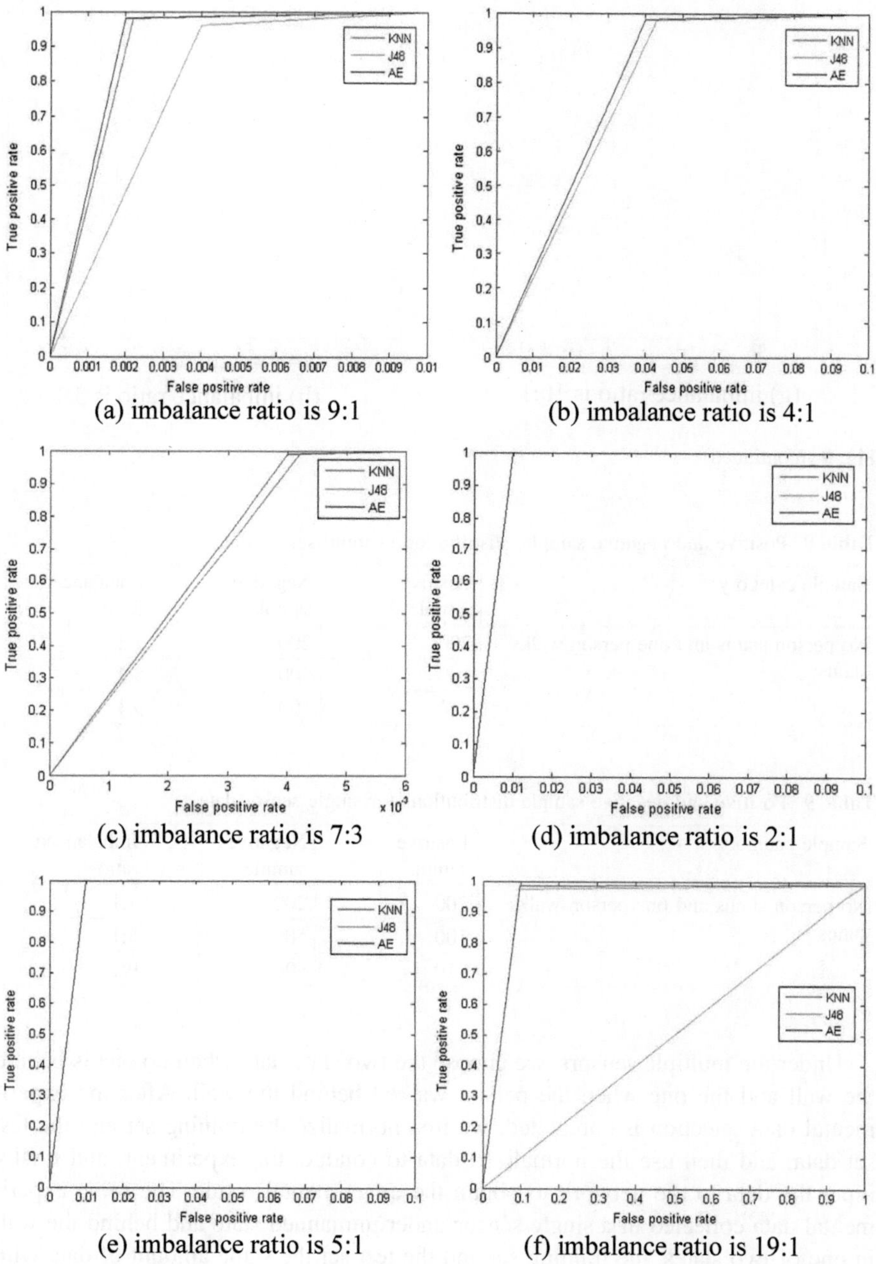

Fig. 9 The ROC curve under different imbalance ratio. Red lines: the ROC curve of KNN algorithm. Green lines: the ROC curve of J48 algorithm. Blue lines: the ROC curve of AE algorithm (Color figure online)

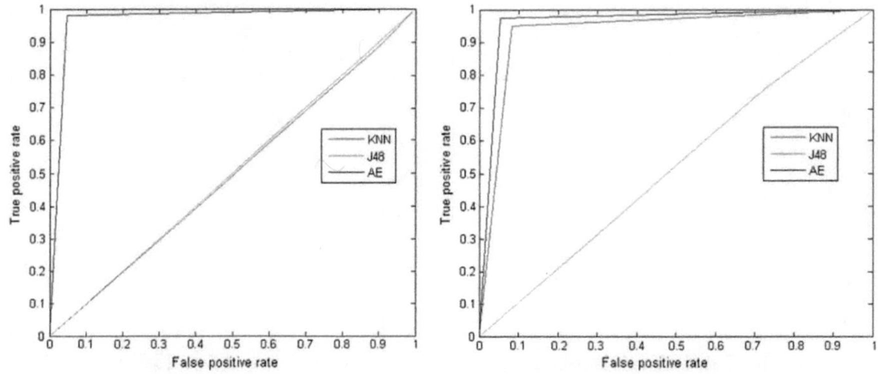

(g) imbalance ratio is 10:1

(h) imbalance ratio is 3:1

Fig. 9 (continued)

Table 8 Positive and negative samples distribution of multi-sensor data

Sample category	Positive sample	Negative sample	Imbalanced Ratio
No person status and one person walks status	200	200	1:1
	100	500	5:1
	10	90	9:1

Table 9 Positive and negative sample distributions for single sensor data

Sample category	Positive sample	Negative sample	Imbalanced ratio
No person status and one person walks status	200	200	1:1
	100	500	5:1
	10	90	9:1

Under the multiple sensors, we choose the two state data when no one is behind the wall and the one when the person walked behind the wall. After the experimental data selection is completed, we first normalize the training set and the test set data, and then use the normalized data to conduct the experiment, and finally input the data to the network to obtain the experimental result. The same experimental data collected in a single sensor under unmanned state and behind the wall in one of two states, the training set and the test set the same amount of data with multiple sensors. We compare the results obtained with the classification of the same two data sets under the same sensor and the same state.

We used the ROC curve to evaluate the experimental results, the closer the ROC curve is to the upper left corner, the higher the accuracy of the classification algorithm. We compared the multi-sensor data results with the single sensor

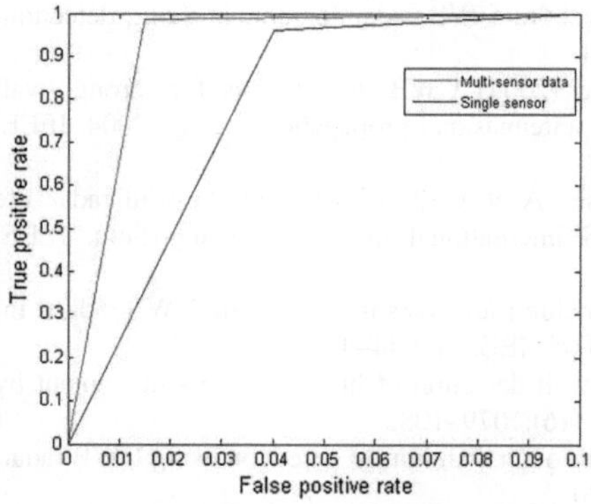

(a) imbalance ratio is 1:1

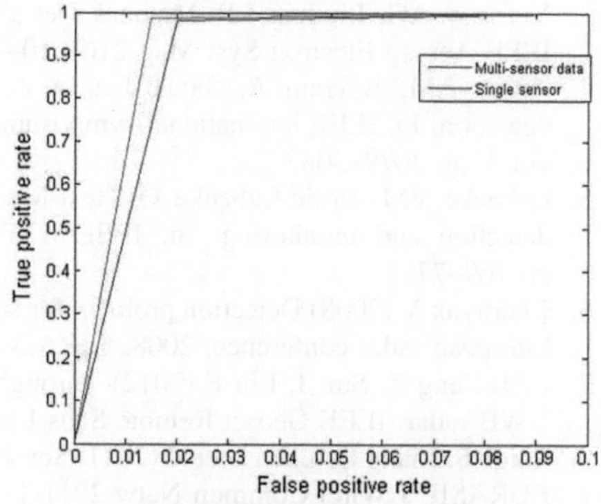

(b) imbalance ratio is 5:1

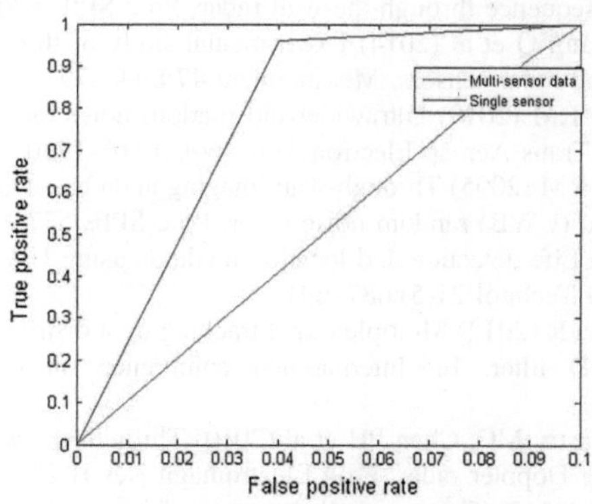

(c) imbalance ratio is 9:1

Fig. 10 The ROC curve under different imbalance ratio. Red lines: the ROC curve of multi-sensor data experiment. Blue lines: the ROC curve of single sensor data experiment (Color figure online)

experiment results. The ROC curves of the two experimental results under different imbalance ratios are shown in Fig. 10. As can be seen from Fig. 10, as the imbalance ratio increases, the experimental results show that the multi-sensor data is better than a single sensor in certain imbalance ratio.

References

1. Kumar A, Liang Q, Li Z, Zhang B, Wu X (2012) Experimental study of through-wall human being detection using ultra-wideband (UWB) radar. Globecom workshops (GC Wkshps), 2012 IEEE
2. Sahu KN, Satyam M, Naidu CD, Sankar KJ (2015) UWB propagation modeling of human being behind a concrete wall for the study of cardiac condition. In: International conference on electrical, 2015, pp 1–5

3. Yarovoy AG, Ligthart LP, Matuzas J et al (2006) UWB radar for human being detection. IEEE Aerosp Electron Syst Mag 21(3):10–14
4. Attiya AM, Bayram A, Safaai-Jazi A et al (2004) UWB applications for through-wall detection. In: IEEE international symposium: antennas and propagation society, 2004. IEEE, vol 3, pp 3079–3082
5. Lubecke VM, Boric-Lubecke O, Host-Madsen A et al (2007) Through-the-wall radar life detection and monitoring. In: IEEE/MTT-S international microwave symposium. IEEE, pp 769–772
6. Chernyak V (2008) Detection problem for searching survivors in rubble with UWB radars. In: European radar conference, 2008. EuRAD 2008. IEEE, pp 44–47
7. Li J, Zeng Z, Sun J, Liu F (2012) Through-wall detection of human being's movement by UWB radar. IEEE Geosci Remote Sens Lett 9(6):1079–1083
8. Singh S, Liang Q, Chen D et al (2011) Sense through wall human detection using UWB radar. EURASIP J Wirel Commun Netw 2011(1):20
9. Zetik R, Crabbe S, Krajnak J et al (2006) Detection and localization of persons behind obstacles using M-sequence through-the-wall radar. Proc SPIE 6201:145–156
10. Kumar A, Li Z, Liang Q et al (2014) Experimental study of through-wall human detection using ultra wideband radar sensors. Measurement 47:869–879
11. Lai CP, Narayanan RM (2010) Ultrawideband random noise radar design for through-wall surveillance. IEEE Trans Aerosp Electron Syst 46(4):1716–1730
12. Lai CP, Narayanan RM (2005) Through-wall imaging and characterization of human activity using ultrawideband (UWB) random noise radar. Proc SPIE 5778:187
13. Sun J, Li M (2011) Life detection and location methods using UWB impulse radar in a coal mine. Int J Min Sci Technol 21(5):687–691
14. Jovanoska S, Thoma R (2012) Multiple target tracking by a distributed UWB sensor network based on the PHD filter. In: International conference on information fusion, 2012, pp 1095–1102
15. Narayanan RM, Shastry MC, Chen PH et al (2010) Through-the-wall detection of stationary human targets using Doppler radar. Prog Electromagn Res B 20:147–166
16. Zhang B, Wang W (2013) Through-wall detection of human being with compressed UWB radar data. EURASIP J Wirel Commun Netw 2013(1):162
17. Levitas B, Matuzas J (2006) UWB radar for human being detection behind the wall. In: International radar symposium, 2006. IRS 2006. IEEE, 2007, pp 85–88
18. Marano S, Gifford WM, Wymeersch H, Win MZ (2010) NLOS identification and mitigation for localization based on UWB experimental data. IEEE J Sel Areas Commun 28 (7):1026–1035
19. Chong CC, Watanabe F, Inamura H (2008) NLOS identification and weighted least-squares localization for UWB systems using multipath channel statistics. EURASIP J Adv Signal Process 2008(1):36
20. Gudmundsson M, El-Kwae EA, Kabuka MR (1998) Edge detection in medical images using a genetic algorithm. IEEE Trans Med Imaging 17(3):469–474
21. Qu W, Wang J, Zheng J, Li G, Teng J (2006) Fuzzy pattern recognition for stress field of box-type steel structure. Earthq Eng Eng Vibr 26(5):177–182
22. Zhao X, Zhang H (2012) Summary: expression classification algorithm and emotional space model. Int J Digit Content Technol Appl 6(3):37–44

Real-Time Wind Velocity Monitoring Based on Acoustic Tomography

Yong Bao and Jiabin Jia

Abstract Wind-related disasters cause tremendous loss around the world, therefore a fast, low-cost but accurate wind velocity monitoring technique is highly desirable and will provide great benefits for wind risk management. Acoustic travel-time tomography, which utilise the dependence of sound speed on the wind velocity along the sound propagation path, is considered to be a promising remote sensing technique for wind velocity monitoring. The success of acoustic tomography technique stems from then various advantage of non-invasive, low cost and easy to implement when compared to other techniques. This chapter describes the fundamentals of the simultaneously multi-channel time-of-flight measurements and the tomographic reconstruction of 2D horizontal wind velocity distributions based on the use of offline iteration method. The feasibility and effectiveness of the proposed methods will be numerically validated in a simulation study.

Keywords Acoustic tomography · Wind velocity · Remote sensing

1 Introduction

Extreme strong wind is dangerous for people and cause significant damage to buildings, infrastructures, agricultural, forestry and property with the damage being highly unpredictable. There are many aspects to be looked into by researchers and engineers [1, 2]. One of the issues is accurate and reliable measurement of wind velocity, particularly in the atmospheric boundary layer, which is the lower part of the atmosphere. A fast, low-cost but accurate wind velocity monitoring technique is highly desirable and will provide great benefits in order to reduce wind-related disasters for safer and securer communities.

Y. Bao · J. Jia (✉)
School of Engineering, Institute for Digital Communications,
The University of Edinburgh, Edinburgh, UK
e-mail: jiabin.jia@ed.ac.uk

T. S. Durrani et al. (eds.), *Geological Disaster Monitoring Based on Sensor Networks*, Springer Natural Hazards, https://doi.org/10.1007/978-981-13-0992-2_9

There are a number of efforts in developing robust and accurate wind velocity measurement techniques. Conventional cup anemometer is the most used device but they have the drawbacks including intrusive, single-point sensing and corrosion in the harsh environment [3]. SODAR (SOund Detection And Ranging) and LIDAR (LIght Detection And Ranging), the two remote sensing technique both employ the Doppler effect to acquire wind observations, are very expensive and relatively power hungry [4]. Compared to other techniques, acoustic tomography is one of few that can deliver accurate quantitative reconstruction of the whole temperature field with lower equipment cost.

Early application of acoustic tomography was firstly proposed in 1990 by Spiesberger [5] and first experimental implementation was completed in 1994 by Wilson and Thomson [6]. Their tomography array consisted of 3 transmitters and 5 receivers to cover a rectangular area of 200 m by 200 m with 50 m resolution. Later, similar acoustic tomography system has been completed in the University of Leipzig, Germany. In serval experiments since 1996 [7–9], the size and number of transducers of the acoustic tomography system were variable, from 100 m by 200 m with 20 m resolution to 200 m by 240 m with 50 m resolution. A first 3D array for acoustic tomography was built at the Boulder Atmospheric Observatory [10]. 9 transmitters and 15 receivers were placed around the square sensing area, with the side length of 80 m, at three adjustable height.

The tomographic reconstruction of the wind velocity is generally difficult. The inverse problem is under-determined due to the limited number of measurements. Therefore, the solution is not unique. Besides, the travel-time measurement is line integral of the wind velocity distribution, which results in an ill-posed inverse problem. This means that the reconstruction results are very sensitive to the measurement noise. Numerous algorithms have been developed to solve the acoustic tomographic inverse problem. These algorithms can be categorised into three main branches including algebraic-based algorithms [8, 11, 12], sparse reconstruction framework [13–15] and stochastic-based algorithms [6, 7, 16–18].

This chapter describes the fundamentals of the measurement method and the tomographic reconstruction of 2D horizontal wind velocity distributions. Section 2.1 reviews the forward modelling of acoustic travel-time tomography. Then in Sect. 2.2 and 2.3 the Time-Of-Flight (TOF) data collection and vector tomography reconstruction algorithm is illustrated. Numerical simulation results are discussed in Sect. 3 and the conclusion is drawn in Sect. 4.

2 Methodology

2.1 Forward Modelling of Acoustic Tomography

Acoustic travel-time tomography mainly utilises the strong dependence of sound propagation on the spatial distribution of air temperature and velocity. A typical setup for acoustic travel-time tomography is illustrated in Fig. 1. The whole square sensing area is surrounded by 16 acoustic transceivers. The time-of-flight (TOF) between each transducer pairs are recorded along 96 ray paths. Given the position of the transceivers and the TOF measurements, the corresponding wind velocity field can be recovered. The acoustic transducer array covers a sensing area of 100 m by 100 m with 5 m resolution, therefore the reconstructed image is a vector field with 20 by 20 pixels.

The wavelength of the acoustic signal is much smaller than the medium inhomogeneities in size, therefore, geometric acoustics model is used and the acoustic signals are considered to propagate along sound ray paths between transducer pairs.

According to the geometrical acoustic assumption, the group velocity $\boldsymbol{u}$ (the observable quantity using acoustic tomography technique) of a sound wave in air is defined as [19]:

$$\boldsymbol{u} = c_L\boldsymbol{n} + \boldsymbol{v} \tag{1}$$

where c_L denotes the Laplace's sound speed which only depends on air temperature, $\boldsymbol{v}$ represents the wind velocity and $\boldsymbol{n}$ is the direction normal to the wave front.

With the use of acoustic transducer array placed around the sensing area, the group velocity is determined according to the time-of-flight (TOF) measurements, which is defined as:

$$\mathrm{T} = \int_{\Gamma} \frac{ds}{(c_L\boldsymbol{n} + \boldsymbol{v})} \tag{2}$$

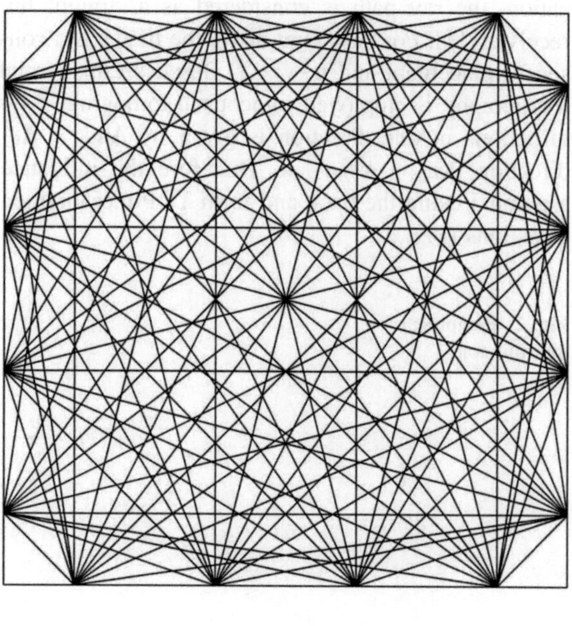

Fig. 1 Acoustic tomography problem setup

where τ is the TOF along ray path Γ and $\boldsymbol{s}$ is the unit direction vector along the ray path.

Equation (2) can be linearized as:

$$\begin{aligned}\tau &= \int_{\Gamma} \frac{ds}{(c_{L0} + \Delta c_L)\boldsymbol{n} + \boldsymbol{v}} \\ &= \int_{\Gamma} \frac{(c_{L0}\boldsymbol{n} - (\Delta c_L \boldsymbol{n} + \boldsymbol{v}))ds}{(c_{L0}\boldsymbol{n})^2 - (\Delta c_L \boldsymbol{n} + \boldsymbol{v})^2} \\ &\approx \int_{\Gamma} \frac{ds}{c_{L0}\boldsymbol{n}} - \int_{\Gamma} \frac{(\Delta c_L \boldsymbol{n} + \boldsymbol{v})ds}{(c_{L0}\boldsymbol{n})^2} \\ &= \tau_0 - \frac{1}{c_{L0}^2} \int_{\Gamma} (\Delta c_L \boldsymbol{n} + \boldsymbol{v})ds\end{aligned} \tag{3}$$

where c_{L0} represents the average Laplace's sound speed and $\Delta c_L = c_L - c_{L0}$ is the group velocity perturbations caused by temperature, and $\boldsymbol{v}$ is the wind velocity. Typically, the group velocity perturbation $\Delta \boldsymbol{u} = \Delta c_L \boldsymbol{n} + \boldsymbol{v}$ is much smaller than c_{L0}, hence Eq. (3) can be rearranged so that TOF perturbations are linearly related with the $\Delta \boldsymbol{u}$.

$$(\tau_0 - \tau)c_{L0}^2 = \int_{\Gamma} (\Delta c_L \boldsymbol{n} + \boldsymbol{v})ds = \int_{\Gamma} \Delta \boldsymbol{u}\, ds \tag{4}$$

Because the acoustic refraction effect is usually ignored in this type of application, the ray path is considered as a straight line from the transmitter to the receiver [11]. For any point along the line, $\boldsymbol{s}$ is a constant vector determined by the sensor placement and $\boldsymbol{n} \cdot \boldsymbol{s} = 1$. Based on this straight line assumption, reciprocal tomography is employed in acoustic tomography system. The group velocity perturbations in one direction is given by $\Delta \boldsymbol{u}_+ = \Delta c_L \boldsymbol{n} - \boldsymbol{v}$, while in the opposite direction $\Delta \boldsymbol{u}_- = \Delta c_L \boldsymbol{n} + \boldsymbol{v}$. Therefore the influence brought by $\Delta c_L \boldsymbol{n}$ and $\boldsymbol{v}$ is separated using the back and forth TOF measurements τ_+ and τ_- in two opposite directions (Fig. 2).

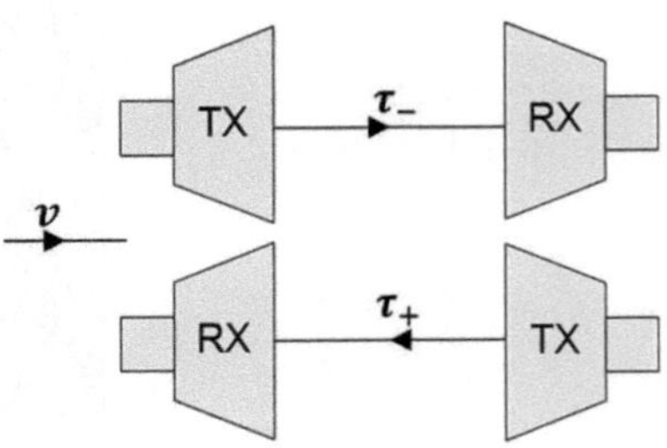

Fig. 2 Reciprocal transmission

$$l_\Gamma = \frac{(\tau_+ - \tau_-)c_{L0}^2}{2} = \int_\Gamma v ds \tag{5}$$

Then Eq. (5) can be discretized as:

$$\begin{bmatrix} lx_{\Gamma,\mathrm{i}} \\ ly_{\Gamma,\mathrm{i}} \end{bmatrix} = \begin{bmatrix} \sum_{j=1}^{N} v_{x_j} s_{i,j} \cos(\theta) \\ \sum_{j=1}^{N} v_{y_j} s_{i,j} \sin(\theta) \end{bmatrix} \tag{6}$$

where v_{x_j} and v_{y_j} are the directional wind velocity in *j*-th pixel, $s_{i,j}$ is the segment length for each ray path cover across one pixel, and $[\cos\theta, \sin\theta]$ is the direction vector from the transmitter to the receiver. Equation (6) can be written in a matrix form:

$$\boldsymbol{L_\Gamma} = \boldsymbol{SV} \tag{7}$$

where $\boldsymbol{L_\Gamma} \in \mathbb{R}^{2N}$ and N is the number of ray path, $\boldsymbol{V} = [\boldsymbol{V_x}; \boldsymbol{V_y}], \boldsymbol{V_x}, \boldsymbol{V_y} \in \mathbb{R}^M$, M is the number of pixels; $\boldsymbol{S} = [\boldsymbol{S_x}; \boldsymbol{S_y}]$, and $\boldsymbol{S_x}, \boldsymbol{S_y} \in \mathbb{R}^{N*M}$ are the directional ray length matrix whose elements are $s_{i,j}\cos(\theta)$ and $s_{i,j}\sin(\theta)$ respectively.

2.2 Acoustic Signal Process for TOF Detection

In conventional acoustic tomography system, each transmitter is sequentially switched on to transmit acoustic signals to the different receivers. That approach has small interference for TOF detection as acoustic signals from different transmitters are separated in a different time slot at the receiver. However, the temporal resolution of the system is limited and it may not be able to capture the dynamic wind velocity changes in the sensing area. In order to improve the temporal resolution of acoustic tomography system, all the acoustic transceivers could transmit and receive acoustic signals simultaneously. As a result, the measurement time is significantly reduced compared with the pairwise sequential measurement strategy. In this operation, each received signal is a summation of all P delayed source signals.

$$y_j(t) = \sum_{i=1}^{\mathrm{P}} x_i(t - \Delta t_{i,j}) + n_j(t), \quad \mathrm{i} \neq \mathrm{j} \tag{8}$$

where $y_j(t)$ refers to the received signal at *j*-th receiver, $x_i(t)$ is the source signal from *i*-th transmitter and $\Delta t_{i,j}$ represents their corresponding delay time, which is the TOF in *i*-th ray path. $n_j(t)$ is the local noise received at *j*-th receiver.

The summed signals must be separated at the receiver side first, then their individual delay time $\Delta t_{i,j}$ can be estimated based on the cross-correlation detection defined as below.

$$\begin{aligned} R_{y_j x_k}(l) &= \sum_{l=-\infty}^{+\infty} \left[y_j(t) x_k(t-l) \right] = \sum_{l=-\infty}^{+\infty} \left[\left[\sum_{i=1, i \neq j}^{K} x_i\left(t - \Delta t_{i,j}\right) + n_j(t) \right] x_k(t-l) \right] \\ &= \sum_{i=1, i \neq j, i \neq \mathrm{k}}^{\mathrm{P}} R_{x_i x_k}\left(l - \Delta t_{i,j}\right) + R_{x_k x_k}\left(l - \Delta t_{k,j}\right) + R_{n_j x_k}(l) \end{aligned} \tag{9}$$

where l denotes the correlation delay and noise term $n_j(t)$ is uncorrelated with the source signal x_i, and x_k is the k-th reference signal waveform for cross-correlation detection.

In order to minimise the interference term $\sum_{i=1, i \neq j, i \neq \mathrm{k}}^{\mathrm{P}} R_{x_i x_k}\left(l - \Delta t_{i,j}\right)$ and noise term $R_{n_j x_k}(l)$, all the source signal waveforms should have good correlation property, such as sharp auto-correlation peak, but low cross-correlation value, for arbitrarily random delay $\Delta t_{i,j}$ and uncorrelated to the noise at the receiver.

Maximum Length Sequence (MLS) has a good asynchronous orthogonal property, which could be used to separate different source signals with arbitrary delay at the receiver [11]. However, when a large number of simultaneous acoustic sources are used, like 16 in this case, the cross-correlation property of MLS is relatively poor. Selecting and combining the preferred pairs of MLS together can improve cross-correlation property, like the Kasami sequence. In fact, the Kasami sequence has near optimal cross-correlation values close to the Welch lower bound [20].

Therefore, the Kasami sequence shown in Fig. 3 is selected to generate acoustic signals for simultaneous source signal transmission and separation. However, it is difficult to practically generate and transmit the Kasami sequence with the acoustic transceivers, because the spectrum of the Kasami sequence is arbitrarily wide with sharp edges and discontinuity in the time domain, but the acoustic transceiver is restricted to a limited bandwidth around a certain frequency. Therefore, before transmission, it is essential to modulate the Kasami sequence with a fixed carrier frequency, then a band-pass filter is applied to smooth the Kasami sequence and control its bandwidth.

As shown in Fig. 4, compared with the original Kasami sequence, the modulated and filtered Kasami sequence has a narrower bandwidth to meet the bandwidth specification for the acoustic transceivers.

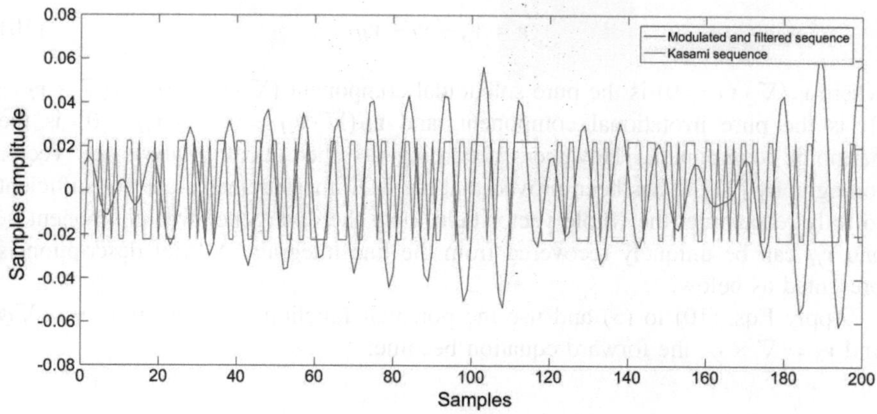

Fig. 3 Kasami sequence (red) and the output signal (blue) in time domain after modulation and band-pass filter (Color figure online)

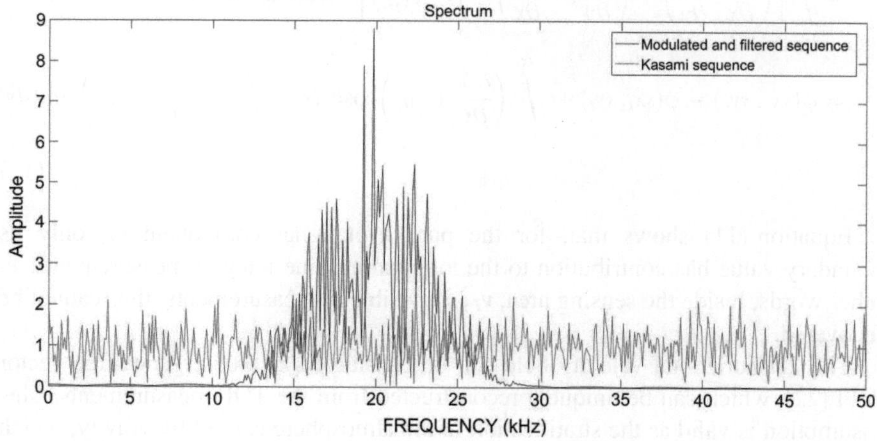

Fig. 4 Kasami sequence (red) and modulated and filtered output sequence (blue) in frequency domain. After modulation and band-pass filter, the bandwidth of output signal is limited and centred at the 20 kHz carrier frequency (Color figure online)

2.3 Wind Velocity Reconstruction

Tomographic reconstruction of wind velocity is considered as a typical vector field tomography problem, which recovers the wind velocity field from its line integrals calculated from reciprocal transmission TOF measurements.

Based on Helmholtz's theorem, a bounded wind velocity field $\boldsymbol{v}$ can be uniquely decomposed into three components.

$$\mathbf{v} = \mathbf{v}_S + \mathbf{v}_I + \mathbf{v}_H \tag{10}$$

where $\mathbf{v}_S(\nabla \cdot \mathbf{v}_S = 0)$ is the pure solenoidal component $(\nabla \cdot \mathbf{v}_S = 0)$, $\mathbf{v}_I(\nabla \times \mathbf{v}_I = 0)$ is the pure irrotational component, and $\mathbf{v}_H(\nabla \cdot \mathbf{v}_H = 0, \nabla \times \mathbf{v}_H = 0)$ is the harmonic component. However, according to theoretical analysis of vector tomography [21], it has been proved that the TOF measurements are not sufficient to fully reconstruct the whole vector field, only the divergence-free component $\mathbf{v}_S$ and $\mathbf{v}_H$ can be uniquely recovered from the line integrals. A brief description is presented as below.

Apply Eqs. (10) to (5) and use the potential function representation $\mathbf{v}_I = \nabla\phi$ and $\mathbf{v}_S = \nabla \times \psi$, the forward equation become:

$$\begin{aligned} l_\Gamma &= \int_{s_t}^{s_r} \mathbf{v}_I + \mathbf{v}_S + \mathbf{v}_H \, ds \\ &= \int_{s_t}^{s_r} \left[\left(\frac{\partial\phi}{\partial x}, \frac{\partial\phi}{\partial y} \right) + \left(\frac{\partial\psi}{\partial y}, -\frac{\partial\psi}{\partial x} \right) + \left(v_{H_x}, v_{H_y} \right) \right] \cdot (\cos\theta \, dx, \sin\theta \, dy) \\ &= \phi(sx_r, sy_r) - \phi(sx_t, sy_t) + \int_{sx_t}^{sx_r} \left(\frac{\partial\psi}{\partial y} + v_{H_x} \right) \cos\theta \, dx + \int_{sy_t}^{sy_r} \left(-\frac{\partial\psi}{\partial x} + v_{H_y} \right) \sin\theta \, dy \end{aligned} \tag{11}$$

Equation (11) shows that, for the pure irrotational component $\mathbf{v}_I$, only its boundary value has contribution to the longitudinal line integral measurements. In other words, inside the sensing area, $\mathbf{v}_I$ is invisible to measurements, thus cannot be recovered.

The 2D horizontal velocity field can be considered as a divergence-free vector field [22], which can be uniquely reconstructed from the TOF measurements. This assumption is valid as the stratification in the atmosphere caused by gravity, which makes the horizontal velocity $\mathbf{v}_{xy}$ greater than the vertical velocity $\mathbf{v}_z$ by a factor of 10–100. Usually for the time-averaged data used here, $\mathbf{v}_z$ can be ignored and therefore $\mathbf{v}_{xy}$ is considered as a divergence-free vector field and the reconstruction of wind velocity field will not be affected by the invisible field problem.

$$\nabla \cdot \mathbf{v}_{xy} = \frac{\partial \mathbf{v}_x}{\partial \mathrm{x}} + \frac{\partial \mathbf{v}_y}{\partial \mathrm{y}} = -\frac{\partial \mathbf{v}_z}{\partial \mathrm{z}} = 0 \tag{12}$$

Among the two divergence-free vector components, v_S and v_H, of the horizontal wind velocity field, it has been proven that, without any other information, both of them can be uniquely recovered from their longitudinal integral measurements [22].

As mentioned before, the forward model describing the relationship between line integral measurements and wind velocity is defined as follow:

$$\boldsymbol{L}_\Gamma = \boldsymbol{SV} \tag{13}$$

The inverse problem for the velocity reconstruction has the general form:

$$\min_{V} \|\boldsymbol{L}_\Gamma - \boldsymbol{SV}\|_2^2 + \eta\|\boldsymbol{DV}\|_2^2 + \alpha\|\boldsymbol{RV}\|_2^2 \tag{14}$$

where the first term $\|\boldsymbol{L}_\Gamma - \boldsymbol{SV}\|_2^2$ is the data fitting term, the second term $\eta\|\boldsymbol{DV}\|_2^2$ is the vector smoothness constraint and $\boldsymbol{D} = [\boldsymbol{D}_x, \boldsymbol{D}_y]$ is the divergence operator. $\boldsymbol{D}_x, \boldsymbol{D}_y \in \mathbb{R}^{N*N}$ are the two directional discrete differentials, which apply discrete 2nd order differential inside the sensing area and 1st order differential for boundary pixel; the third term $\alpha\|\boldsymbol{RV}\|_2^2$ is the vector smoothness constraints, and $\boldsymbol{R}$ is the vector Laplace operator $\boldsymbol{RV} = \nabla^2\boldsymbol{v}$, which can be built using $\boldsymbol{D}_x, \boldsymbol{D}_y$. Details are showed as below:

$$\begin{aligned}
\boldsymbol{RV} &= \nabla^2\boldsymbol{v} = \left(\nabla^2\boldsymbol{v}_x, \nabla^2\boldsymbol{v}_y\right) \\
&= \left(\frac{\partial^2\boldsymbol{v}_x}{\partial x^2} + \frac{\partial^2\boldsymbol{v}_x}{\partial y^2}, \frac{\partial^2\boldsymbol{v}_y}{\partial x^2} + \frac{\partial^2\boldsymbol{v}_y}{\partial y^2}\right) \\
&= \left(\left(\partial_{xx}^2 + \partial_{yy}^2\right)\boldsymbol{v}_x, \left(\partial_{xx}^2 + \partial_{yy}^2\right)\boldsymbol{v}_y\right) \\
&= \left[\boldsymbol{D}_x\boldsymbol{D}_x + \boldsymbol{D}_y\boldsymbol{D}_y, \boldsymbol{D}_x\boldsymbol{D}_x + \boldsymbol{D}_y\boldsymbol{D}_y\right]\boldsymbol{V}
\end{aligned} \tag{15}$$

Therefore $\boldsymbol{R} = \left[\boldsymbol{D}_x\boldsymbol{D}_x + \boldsymbol{D}_y\boldsymbol{D}_y, \boldsymbol{D}_x\boldsymbol{D}_x + \boldsymbol{D}_y\boldsymbol{D}_y\right]$.

To solve the inverse problem defined by Eq. (15), an upgrade of Simultaneous Iterative Reconstruction Technique (SIRT) algorithm is applied [9]. This algorithm has the advantage of fast convergence and stability. Generally, it is defined as follow:

$$\boldsymbol{V}^{(k+1)} = \boldsymbol{V}^{(k)} + \lambda\boldsymbol{P}\left[\boldsymbol{S}^T\boldsymbol{W}\left(\boldsymbol{L}_\Gamma - \boldsymbol{SV}^{(k)}\right) - \alpha\boldsymbol{R}^T\boldsymbol{RV}^{(k)} - \eta\boldsymbol{D}^T\boldsymbol{DV}^{(k)}\right] \tag{16}$$

$$\boldsymbol{P} = \mathrm{diag}(1/LP_1, 1/LP_2, \ldots, 1/LP_{2N}) \tag{17}$$

$$\boldsymbol{W} = \mathrm{diag}(1/LR_1, 1/LR_2, \ldots, 1/LR_{2M}) \tag{18}$$

where λ is the constant iteration step size, α and η are the regularization parameters, k in this section denotes the iteration for linear reconstruction, $LP_j = \sum_{i=1}^{2M} s_{i,j}$ is the directional segment length for all the ray paths across j-th pixel and $\boldsymbol{P}$ is the diagonal preconditioner which is used to geometrically weight the least square cost function and therefore ensure a better noise tolerance and prevent semi-convergence for reconstruction. $LR_i = \sum_{j=1}^{2N} s_{i,j}$ is the length of i-th ray path and $\boldsymbol{W}$ is the normalised operator to make sure that the sum of each row of S equals to 1. By weighting with $\boldsymbol{W}$, when residual norm is minimised, the rays that intersect larger portions of image can tolerate larger errors than these are much shorter [23]. The iteration step size, λ is chosen so that convergence condition holds

$$0<\varepsilon<\lambda<\rho \tag{19}$$

where ρ is the matrix radius of $(\boldsymbol{PS^T WS} + \boldsymbol{P}(\boldsymbol{R^T R}+\boldsymbol{D^T D}))$.

In most real-time monitoring system, non-iterative methods are preferred for fast reconstruction speed. To improve the accuracy of non-iterative method, many researchers tried to iteratively calculate the optimal inversion operator beforehand for non-iterative online reconstruction, for instance, Offline Iteration Online Reconstruction (OIOR) [24] based on Landweber iteration, and Direct Landweber (DLW) based on modified Landweber [25]. In order to build a real-time acoustic tomography system, the offline iteration method is applied based on the SIRT method and consequently, the reconstruction time can be reduced to the same level as non-iterative method.

The principle of this method is to design an iteration method, whose iteration procedure is linear and independent of measurement data, which means that the iteration of Eq. (16) can be rewritten as

$$\boldsymbol{V}^{(k+1)} = \boldsymbol{B}\boldsymbol{V}^{(k)} + \boldsymbol{G}\boldsymbol{L}_\Gamma \tag{20}$$

where $\mathbf{B} = (\boldsymbol{I} - \lambda\boldsymbol{PS^T WS} - \lambda\alpha\boldsymbol{R^T R} - \lambda\eta\boldsymbol{D^T D})$ and $\mathbf{G} = \lambda\boldsymbol{PS^T W}$

The solution $\boldsymbol{V}^{k+1}$ is decomposed into two parts, the iterative term $\boldsymbol{C}^{k+1}$ and the non-iterative term $\boldsymbol{\varepsilon}$

$$\boldsymbol{V}^{k+1} = \boldsymbol{C}^{k+1}\boldsymbol{\varepsilon} \tag{21}$$

Substituting Eq. (21) into Eq. (20)

$$\boldsymbol{C}^{k+1}\boldsymbol{\varepsilon} = \boldsymbol{B}\boldsymbol{C}^{k}\boldsymbol{\varepsilon} + \boldsymbol{G}\boldsymbol{L}_\Gamma \tag{22}$$

Let $\boldsymbol{\varepsilon} = \boldsymbol{L}_\Gamma$, then the iterative term $\boldsymbol{C}^{k+1}$ is independent of TOF measurement $\boldsymbol{\tau}$ and can be calculated offline in advance using equation below.

$$\boldsymbol{C}^{k+1} = \boldsymbol{B}\boldsymbol{C}^{k} + \boldsymbol{G} \tag{23}$$

Given the iteration number k, after $\boldsymbol{C}$ is iteratively obtained offline and all the measurements L_Γ are measured, the wind velocity $\boldsymbol{V}$ can be determined by Eq. (24).

$$\boldsymbol{V} = \boldsymbol{C}\boldsymbol{L}_\Gamma \tag{24}$$

For the offline iteration calculation of $\boldsymbol{C}$, its iteration number k need to be determined beforehand.

3 Simulation Results and Discussions

3.1 TOF Measurement

For the purpose of real-time measurement, the acoustic signal waveform is designed based on the modulated and filtered Kasami sequence. Table 1 illustrates the parameters of generating the waveform for the acoustic source signal. Cross-correlation detection is used to estimate the TOF. In the simulation, all the transmitted signals reached the receivers with preset delay. The cross-correlation detection of TOFs is shown in the Fig. 5, where the arrival time of the transmitted signal from the transceiver 1 to the transceiver 3, 5, 7 and 9 are indicated on the cross-correlation peaks. The overlaps between preset delay and detected delay demonstrate very good accuracy of TOF detection.

Generally, compared with the pairwise TOF measurement process, the total measurement time per frame is reduced from 16 to 1 s.

3.2 Wind Velocity Reconstruction

The performance of proposed vector field reconstruction algorithm is tested on three typical vector field phantoms, including a pure solenoidal vector field $\boldsymbol{v}_S$, a divergence-free vector field which contains the harmonic component $\boldsymbol{v}_S + \boldsymbol{v}_H$, and a composite $\boldsymbol{v}_S + \boldsymbol{v}_H + \boldsymbol{v}_I$ vector field. The last one contains the irrotational vector component when the vertical wind velocity cannot be ignored, despite it is much smaller than the solenoidal component by a factor of 10. To create these phantoms, three fundamental velocity components, solenoidal component $\boldsymbol{v}_S$, irrotational component $\boldsymbol{v}_I$ and harmonic component $\boldsymbol{v}_H$ are created in Fig. 6.

In the simulation, all the TOF measurements contain white Gaussian noise with 40 dB SNR. The algorithm parameters, such as the weight parameter α and η for the divergence-free regularization and vector Laplacian regularization, are empirically determined based on a series of practices and the same parameters are applied

Table 1 Parameters of acoustic waveform

Sampling frequency	200 kHz
Carrier frequency	20 kHz
Band-pass filter bandwidth	10 kHz
Filter length	81
Filter window	Kaiser
Total measurement time per frame	1 s
Pulse duration	0.01 s
Kasami code polynomial	[14, 13, 8, 4, 0]

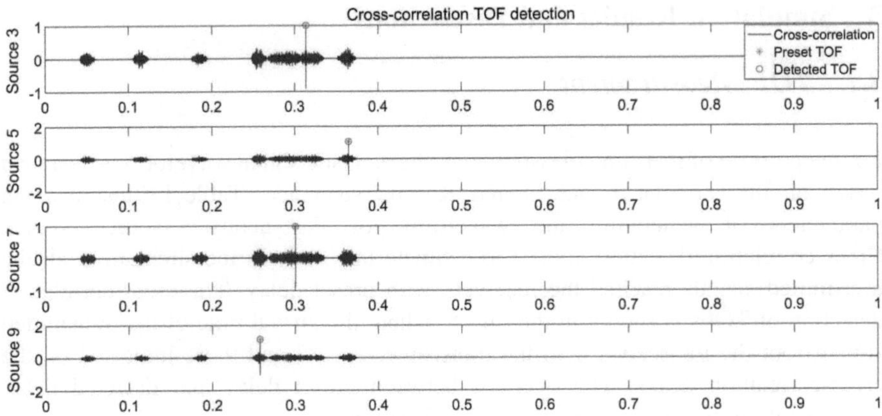

Fig. 5 Cross-correlation TOF detection results from simulation. The preset delay is marked in green asterisks and the detected delay is marked in the red circles (Color figure online)

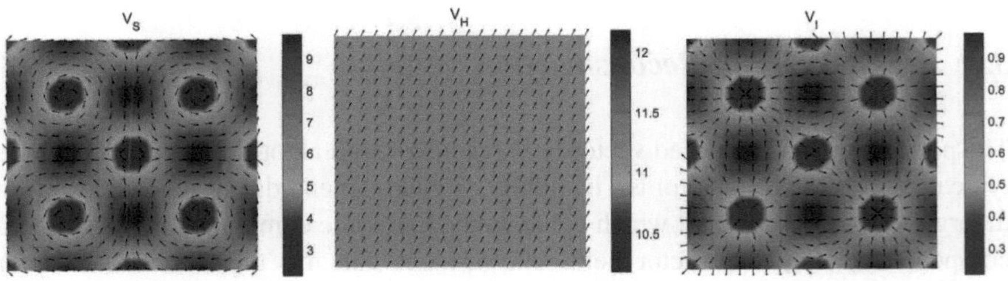

Fig. 6 Three vector field components, from left to right, the solenoidal vector field v_s, the harmonic vector field v_H and the irrotational vector field v_I

to the test phantoms, which is 0.01 and 0.0001 respectively. When implementing the offline iteration algorithm, the pre-calculation iteration number is set to be 200 and the iteration step size is 0.4, which is sufficient for convergence at 40 dB SNR and regularization condition.

To quantitatively evaluate the accuracy of the reconstructed vector field, the relative image error between the reconstructed wind velocity and the true phantom is employed, which is defined as follow:

$$E_{image} = \frac{\|\boldsymbol{V}_{re}(i) - \boldsymbol{V}_{ori}(i)\|_2^2}{\|\boldsymbol{V}_{ori}(i)\|_2^2} \tag{25}$$

Figure 7 illustrates the wind velocity reconstruction results and the relative image errors. The first phantom is designed to evaluate the reconstruction accuracy on the pure solenoidal vector field, where the velocity vanishes at the boundary.

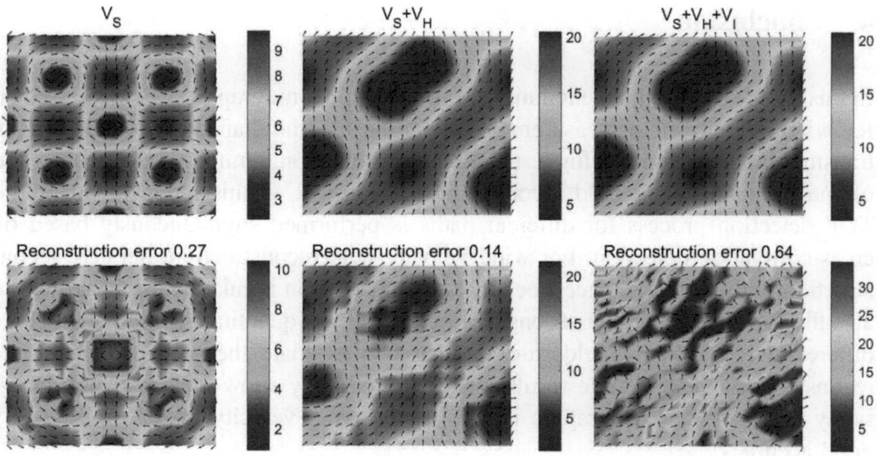

Fig. 7 Simulation phantoms and reconstructed wind velocity fields. The arrows represent the directions of the wind field and the colours indicate its amplitude (Color figure online)

The reconstruction result correctly shows the vortex shape and center positions. There seems to appear some artefact and small discontinuities on the magnitude of the velocity field. The relative image reconstruction error is 0.26, which might be related to the modelling error due to the discretization of the domain and the limited measurement rays paths (96 ray path for 16 transceivers).

The second phantom is designed to evaluate the reconstruction accuracy for the source-free vector field with non-zero boundary velocity, which is a more general case. The velocity field contains the harmonic components, which is both source-free and curl-free, and curl-free means that it also has an invisible field problem. Norton has concluded that measurements of the normal velocity on the boundary can be used to resolve the ambiguity of the harmonic part, and then v_S and v_H can be recovered separately [21]. Later, Ivana further proved that no extra measurements are needed to determine the source-free vector field $v_S + v_H$. Because the entire vector field needs to be reconstructed, it is not necessary to sperate two components [22]. The relative reconstruction error of the second phantom is 0.14, which is lower than the error of the first case. This is due to the low complexity of harmonic component.

The third phantom tests the velocity field when the irrotational component cannot be ignored. The relative image error increases to 0.64 in this case. The reconstruction accuracy is largely affected by the irrotational component, even the source-free component is much larger than the irrotational vector field by a factor of 10. This phantom shows that acoustic travel-time tomography is more suitable for the horizontal slice of wind velocity in the stratified atmosphere, where the wind velocity is a 2D source-free vector field.

4 Conclusion

In this chapter, the fundamental methods of acoustic tomography system are studied for wind velocity field measurement, including the simultaneous acoustics signal transmission and time-of-flight (TOF) collection along multiple ray paths, and offline iteration vector field reconstruction algorithm. In this mode of operation, TOF detection process for different paths is performed simultaneously based on cross-correlation detection. For wind velocity field reconstruction, iterative reconstruction with the divergence-free and vector Laplacian regularization is applied in an offline mode for online reconstruction with good quantitative accuracy. Three different wind velocity fields are simulated to evaluate the performance of the reconstruction method. The results of feasibility study show that acoustic tomography can provide tomographic images of the wind velocity field in a relatively good accuracy.

References

1. Marchigiani R et al (2013) Wind disasters: a comprehensive review of current management strategies. Int J Crit Illn Inj Sci 3(2):130–142
2. Tamura Y (2009) Wind induced damage to buildings and disaster risk reduction. In: Proceedings of the APCWE-VII, Taipei, Taiwan
3. Mikkelsen T (2010) Remote sensing of wind. Remote sensing for wind energy, pp 7–20
4. Sheh R et al (2006) A low-cost, compact, lightweight 3d range sensor. In: Australian conference on robotics and automation
5. Spiesberger JL, Fristrup KM (1990) Passive localization of calling animals and sensing of their acoustic environment using acoustic tomography. Am Nat 135:107–153
6. Keith Wilson D, Thomson DW (1994) Acoustic tomographic monitoring of the atmospheric surface layer. J Atmos Oceanic Technol 11(3):751–769
7. Vecherin SN et al (2007) Tomographic reconstruction of atmospheric turbulence with the use of time-dependent stochastic inversion. J Acoust Soc Am 122(3):1416–1425
8. Ziemann A, Arnold K, Raabe A (2002) Acoustic tomography as a remote sensing method to investigate the near-surface atmospheric boundary layer in comparison with in situ measurements. J Atmos Oceanic Technol 19(8):1208–1215
9. Ziemann A, Arnold K, Raabe A (1998) Acoustic tomography in the atmospheric surface layer. In: Annales Geophysicae. Springer, Berlin
10. Ostashev V et al (2008) Recent progress in acoustic tomography of the atmosphere. In: IOP conference series: earth and environmental science. IOP Publishing
11. Holstein P et al (2004) Acoustic tomography on the basis of travel-time measurement. Meas Sci Technol 15(7):1420
12. Jovanovic I (2008) Inverse problems in acoustic tomography
13. Jovanovic I et al (2007) Efficient and stable acoustic tomography using sparse reconstruction methods. In: 19th international congress on acoustics
14. Dogan Z et al (2012) 3D reconstruction of wave-propagated point sources from boundary measurements using joint sparsity and finite rate of innovation. In: 2012 9th IEEE international symposium on biomedical imaging (ISBI). IEEE
15. Toši I et al (2010) Ultrasound tomography with learned dictionaries. In: 2010 IEEE international conference on acoustics, speech and signal processing. IEEE

16. Vecherin SN et al (2006) Time-dependent stochastic inversion in acoustic travel-time tomography of the atmosphere. J Acoust Soc Am 119(5):2579–2588
17. Liu Y et al (2015) A method for simultaneous reconstruction of temperature and concentration distribution in gas mixtures based on acoustic tomography. Acoust Phys 61(5):597–605
18. Kolouri S, Azimi-Sadjadi MR, Ziemann A (2014) Acoustic tomography of the atmosphere using unscented Kalman filter. IEEE Trans Geosci Remote Sens 52(4):2159–2171
19. Ostashev VE, Wilson DK (2015) Acoustics in moving inhomogeneous media. CRC Press, Boca Raton
20. Li H, Takata S, Yamada A (2011) Tomographic measurement of vortex air flow field using multichannel transmission and reception of coded acoustic wave signals. Jpn J Appl Phys 50 (7S):07HC09
21. Norton SJ (1992) Unique tomographic reconstruction of vector fields using boundary data. IEEE Trans Image Process 1(3):406–412
22. Jovanovic I, Sbaiz L, Vetterli M (2009) Acoustic tomography for scalar and vector fields: theory and application to temperature and wind estimation. J Atmos Oceanic Technol 26 (8):1475–1492
23. Gregor J, Fessler JA (2015) Comparison of SIRT and SQS for regularized weighted least squares image reconstruction. IEEE Trans Comput Imaging 1(1):44–55
24. Liu S et al (2004) Prior-online iteration for image reconstruction with electrical capacitance tomography. IEE Proc Sci Meas Technol 151(3):195–200
25. Kim BS et al (2015) Electrical resistance imaging of two-phase flow using direct Landweber method. Flow Meas Instrum 41:41–49

Joint Optimization of Resource Allocation with Inter-beam Interference for a Multi-beam Satellite and Terrestrial Communication System

Min Jia, Ximu Zhang, Qing Guo and Xuemai Gu

Abstract Satellite plays an important role in the environment sensing during the emergency scenario, especially to robust to the natural disaster. Integrated satellite-terrestrial network which combines both advantages of the satellite network and the terrestrial network can achieve all-day seamless coverage and broad coverage areas. Integrated satellite and terrestrial networks can be used to solve communication problems in natural disasters, forestry monitoring and control, and military communication. Unlike traditional communication methods, integrated networks are effective solutions because of their advantages in communication, remote sensing, monitoring, navigation, and all-weather seamless coverage. Monitoring, urban management, and other aspects will also have a wide range of applications. The multi-beam satellite communication system increases spectrum utilization significantly, but it can cause satellite network and the terrestrial network intensive co-frequency interference. However, the exclusion zone makes the signal to interference and noise ratio (SINR) of the satellite-terrestrial link increase significantly. In this chapter, we propose a first builds an integrated network overlay model, and divides the satellite network into two categories: terrestrial network end users and satellite network end users. The energy efficiency, throughput, and signal-to-noise ratio (SINR) are deduced and analyzed. In this chapter, we discuss the influence of various factors, such as transmit power, number of users, size of the protected area, and terminal position, on energy efficiency and SINR. A satellite-sharing scheme with a combination of the user location and a protection zone with high energy efficiency and anti-jamming capability are proposed, to provide better communication quality for end users in integrated satellite and terrestrial networks. Finally, the numerical results show that the performance of proposed scheme is superior to the traditional integrated satellite-terrestrial scheme.

M. Jia (✉) · X. Zhang · Q. Guo · X. Gu
School of Electronics and Information Engineering, Communication Research Center, Harbin Institute of Technology, Room 1123, Yikuang Street No. 2, Science Park of Harbin Institute of Technology, Harbin City, Heilongjiang Province, China
e-mail: jiamin@hit.edu.cn
URL: http://homepage.hit.edu.cn/pages/jiamin

T. S. Durrani et al. (eds.), *Geological Disaster Monitoring Based on Sensor Networks*, Springer Natural Hazards, https://doi.org/10.1007/978-981-13-0992-2_10

Keywords Integrated satellite-terrestrial network · Spectrum sharing scheme Energy efficiency · Inter-cell fairness · Throughput

1 Introduction

Integrated satellite-terrestrial network is the significant information bridge of "the Belt and Road". Satellite network and terrestrial network coordinate with each other, which serves the terminal users. The company such as 7-ELEVEN INC has been working on building a satellite-based integrated network based on ancillary terrestrial components (ATC) [1]. In 2009, Japanese scholars proposed to build a satellite terrestrial integrated mobile communication system (STICS) for emergency support [2]. The satellite air interface technology based on LTE were submitted to ITU in 2012 [3] and the National High Technology Research and Development Program ("863"Program) of China proposed the integration information network of the space air terrestrial integrated information network, which accelerated the process of satellite access to the fifth generation mobile communication [4, 5].

As the satellite spectrum sharing scheme is proposed, spectrum utilization rate increased significantly, but the satellite network and the terrestrial network will have intensive co-frequency interference. Considering beam-edge user (BEU) with severe interference, Vincent Deslandes divided user terminals into two main categories: satellite network terminals and terrestrial base station users and then proposed a scheme of integrated satellite-terrestrial network spectrum sharing based on exclusion zone in 2010 [6]. Integrated satellite-terrestrial network spectrum sharing based on exclusion zone enhance the SINR, however, system throughput is reduced. Recently, the concept of integrated satellite-terrestrial soft frequency reuse has been investigated. Specifically, [7, 8] verified that the throughput of the scheme satellite-terrestrial soft frequency reuse had increased. Although some outstanding works have been dedicated in studying satellite spectrum sharing scheme [9, 10], energy efficiency of integrated satellite-terrestrial network system is relatively low. However, high energy efficiency conforms to the expectations of green communications for the fifth generation mobile communications [11, 12]. Literature [13, 14] optimized energy efficiency, however ignored the inter-cell fairness and throughput. Considering the throughput, energy efficiency, SINR and inter-cell fairness of the integrated satellite-terrestrial spectrum sharing scheme has not been proposed yet, and this motivates our current work.

In this chapter, we present energy-efficient integrated satellite-terrestrial spectrum sharing scheme based on inter-cell Fairness. Firstly, we established satellite beam, macro base station (MBS) and remote radio heads (RRHs) three layers covered scene, and perform cover analysis [15]. Then, we establish exclusion zone and sort the degree of frequency isolation. An energy-efficient integrated satellite-terrestrial spectrum sharing scheme based on inter-cell fairness is proposed according to user density and degree of frequency isolation. The simulation results show that the scheme we propose has high energy efficiency, SINR, inter-cell fairness and system throughput under different user density.

2 System Model and Interference Analysis

2.1 System Model

This chapter consider satellite beams, macro base station (MBS), remote radio heads (RRHs) three layers covered scene as shown in Fig. 1.

For further consideration, it is assumed that for a terminal user where it is located in a satellite beam, the received interference signal can be written as

$$I_{u(i,k)} = I_{\mathrm{t},u(i,k)} + I_{\mathrm{sat},u(i,k)} \tag{1}$$

where $I_{\mathrm{sat},u(i,k)}$ is the interference of satellite to $u(i,k)$, $I_{\mathrm{t},u(i,k)}$ is the interference of MBS/RRH to $u(i,k)$. In integrated satellite-terrestrial network, satellite beam, macro base station (MBS) and remote radio heads (RRHs) three layers covered caused inter-layer interference seriously. This will affect the system throughput, effective coverage, SNR, energy efficiency and other properties, thus we reduced intra-tier and inter-tier interference, respectively. Through the airspace multi-point cooperative transmission to reduce $I_{\mathrm{t},u(i,k)}$, so as to achieve the purpose of

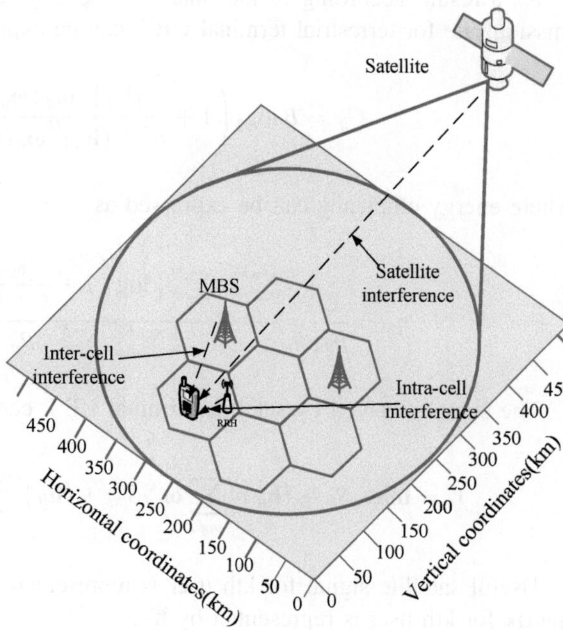

Fig. 1 Three layers covered scene of integrated satellite-terrestrial network

eliminating intra-tier interference $I_{u(i,k)}$. Assuming that there is a satellite cell, an HPN, and fth RRHs. We can define the fth RRH has K_f th terminals where f = {0, 1, 2, …, F}, the index f = 0 refers to the MBS.

The signal received by terrestrial terminal UE k can be written as

$$Y_i = (\mathbf{h}_{if})^H \boldsymbol{\omega}_{if} S_{if} + (\mathbf{h}_{if})^H \sum_{m \neq i} \boldsymbol{\omega}_{mf} S_{mf} + (\mathbf{h}_{if})^H \sum_{l \neq f} \sum_{m=1}^{N} \boldsymbol{\omega}_{ml} S_{ml} + \mathbf{h}_{is} \boldsymbol{\omega}_{is} S_{is} + n_{if} \tag{2}$$

where RRHs/MBS downlink channel matrix for kth user is represented by $\mathbf{h}_{if}$. RRHs/MBS downlink beamforming vector for kth user is represented by $\boldsymbol{\omega}_{if}$, where the index f = 0 refers to the MBS downlink beamforming vector and f = {1, 2, ……, N} denote the set of RRHs downlink beamforming vector. S_{if} is the useful signal received by the terrestrial RRH/MBS. It is assumed that the scalar-valued data stream S_{if} is temporally white with zero mean and unit variance. Satellite signal for kth user is represented by S_{is} and downlink channel matrix for kth user is represented by $\mathbf{h}_{is}$.

As the interlayer interference intense, the use of Coordinated Multiple Points Transmission/Reception(CoMP) is necessary. The inter-layer interference and the intra-layer interference are coordinated by CoMP, efficient and effective zero-forcing (ZF) is used as an example to precoding. $\mathbf{h}_{if}$ and $\boldsymbol{\omega}_{if}$ in the above formula meet

$$(\mathbf{h}_{if})^H \boldsymbol{\omega}_{ml} = 0, \quad \forall m \neq iORl \neq f \tag{3}$$

As a result, according to the Shannon capacity formula, the achievable transmission rate for terrestrial terminal UE k can be expressed as

$$C_{if} = B \log_2 \left(1 + \frac{(\mathbf{h}_{if})^H \boldsymbol{\omega}_{if} (\boldsymbol{\omega}_{if})^H \mathbf{h}_{if}}{\sigma^2 + (\mathbf{h}_{is})^H \boldsymbol{\omega}_{is} (\boldsymbol{\omega}_{is})^H \mathbf{h}_{is}} \right) \tag{4}$$

where energy efficiency can be expressed as

$$\eta_{if} = \frac{C}{p_{\mathrm{tot}}} = \frac{\sum_{f=0}^{F} \sum_{i=1}^{N} \log \left(1 + \frac{(\mathbf{h}_{if})^H \boldsymbol{\omega}_{if} (\boldsymbol{\omega}_{if})^H \mathbf{h}_{if}}{\sigma^2 + (\mathbf{h}_{is})^H \boldsymbol{\omega}_{is} (\boldsymbol{\omega}_{is})^H \mathbf{h}_{is}} \right)}{\xi \sum_{f=0}^{F} \sum_{i=1}^{N} P_{if} + P_{\mathrm{C}}} \tag{5}$$

The signal received by satellite terminal UE k can be written as

$$Y_i = \mathbf{h}_{is} \boldsymbol{\omega}_{is} S_{is} + (\mathbf{h}_{if})^H \sum_{m \neq i} \boldsymbol{\omega}_{mf} S_{mf} + (\mathbf{h}_{if})^H \sum_{l \neq f} \sum_{m=1}^{N} \boldsymbol{\omega}_{ml} S_{ml} + n_{if} \tag{6}$$

Useful satellite signal for kth user is represented by S_{is} and downlink channel matrix for kth user is represented by $\mathbf{h}_{is}$.

Similarly, the inter-layer interference and the intra-layer interference are coordinated by CoMP, efficient and effective zero-forcing (ZF) is used to precoding $\mathbf{h}_{if}$ and $\boldsymbol{\omega}_{if}$ in the above formula meet

$$(\mathbf{h}_{if})^H \boldsymbol{\omega}_{ml} = 0, \quad \forall m \neq iORl \neq f \tag{7}$$

According to the Shannon capacity formula, the achievable transmission rate for satellite terminal UE k can be expressed as

$$C_{\text{is}} = B\log_2\left(1 + \frac{(\mathbf{h}_{is})^H \boldsymbol{\omega}_{is}(\boldsymbol{\omega}_{is})^H \mathbf{h}_{is}}{\sigma^2}\right) \tag{8}$$

where energy efficiency for satellite terminal UE k can be expressed as

$$\eta_{is} = \frac{C}{p_{\text{tot}}} = \frac{\sum_{f=0}^{F}\sum_{i=1}^{N}\log\left(1 + \frac{(\mathbf{h}_{is})^H \boldsymbol{\omega}_{is}(\boldsymbol{\omega}_{is})^H \mathbf{h}_{is}}{\sigma^2}\right)}{\xi\sum_{f=0}^{F}\sum_{i=1}^{N}P_{is} + P_{\text{C}}} \tag{9}$$

Total throughput of the integrated satellite-terrestrial network system can be expressed as

$$C = B\sum_{f=0}^{F}\sum_{i=1}^{N}\log_2\left(1 + \frac{(\mathbf{h}_{if})^H \boldsymbol{\omega}_{if}(\boldsymbol{\omega}_{if})^H \mathbf{h}_{if}}{\sigma^2 + (\mathbf{h}_{is})^H \boldsymbol{\omega}_{is}(\boldsymbol{\omega}_{is})^H \mathbf{h}_{is}}\right)\left(1 + \frac{(\mathbf{h}_{is})^H \boldsymbol{\omega}_{is}(\boldsymbol{\omega}_{is})^H \mathbf{h}_{is}}{\sigma^2}\right) \tag{10}$$

Energy efficiency of the integrated satellite-terrestrial network system can be expressed as

$$\eta_{\text{EE}} = \frac{C}{p_{\text{tot}}} = \frac{B\sum_{f=0}^{F}\sum_{i=1}^{N}\log_2\left(1 + \frac{(\mathbf{h}_{if})^H \boldsymbol{\omega}_{if}(\boldsymbol{\omega}_{if})^H \mathbf{h}_{if}}{\sigma^2 + (\mathbf{h}_{is})^H \boldsymbol{\omega}_{is}(\boldsymbol{\omega}_{is})^H \mathbf{h}_{is}}\right)\left(1 + \frac{(\mathbf{h}_{is})^H \boldsymbol{\omega}_{is}(\boldsymbol{\omega}_{is})^H \mathbf{h}_{is}}{\sigma^2}\right)}{\xi\sum_{f=0}^{F}\sum_{i=1}^{N}P_{is} + \xi\sum_{f=0}^{F}\sum_{i=1}^{N}P_{if} + P_{\text{C}}} \tag{11}$$

2.2 Interference Analysis

Integrated satellite-terrestrial network users can be divided into two categories: satellite network terminals and terrestrial network terminals.

The terrestrial base station signal propagation model [14] in the macro cell can be expressed as

$$PL = 128.1 + 37.6\ \log_{10}(R) + \log_{10}(F) \tag{12}$$

where R denotes the distance from the terminal to the base station, $\log_{10}(F)$ denotes the shadow fading distributed normally.

The link loss of satellite signals only consider the fading of free space, regardless of the influence of rain and other weather and can be expressed as

$$PL = 92.45 + 20\log_{10}f + 20\log_{10}d \tag{13}$$

It is calculated that the interference of the terrestrial terminal downlink to the satellite network terminal downlink is negligible for the satellite network terminal. On the contrary, the terrestrial network terminal uplink signal to the satellite network terminal uplink interference is serious. Because the terrestrial network terminal and satellite network terminal transmission power are almost same. In terms of free space loss, the satellite is particular far from the terrestrial, free space loss of terrestrial network terminal and satellite network terminal is very small. The environment in which the terminal is located will affect the degree of such interference. Even if the terrestrial network terminal is in the urban area, the shadow effect can weaken interference to the satellite uplink, but still can not avoid some terminal and satellite existence direct path.

For the terrestrial network terminal, interference of the satellite network terminal uplink signal is similar to the interference of terrestrial base station downlink signal. In addition, the transmitting power of the satellite terminal is higher than that of the terrestrial. Therefore, the coexistence of interference in this scenario is smaller than that of the terrestrial base station downlink to the satellite network terminal downlink, which is also negligible. In contrast, the interference of the satellite downlink signal to the terrestrial network terminal uplink is serious. The main influencing factors are the size of the protected area and whether the terrestrial network terminal is a cell center user or an edge user.

3 Proposed Integrated Satellite-Terrestrial Cognitive Spectrum Sharing Scheme

3.1 Traditional Integrated Satellite-Terrestrial Spectrum Sharing Scheme

The whole spectrum of the satellite integrated network is divided into seven parts, seven bands are a cluster, and the frequency reuse factor is 7 [16, 17]. Satellite - terrestrial spectrum 7-color multiplex diagram is shown in Fig. 2.

In Fig. 2, terrestrial cell can reuse all frequency with only this satellite beam frequency band exception in the traditional satellite spectrum sharing scheme. This make the limited tight spectrum of resources can effectively improve the overall

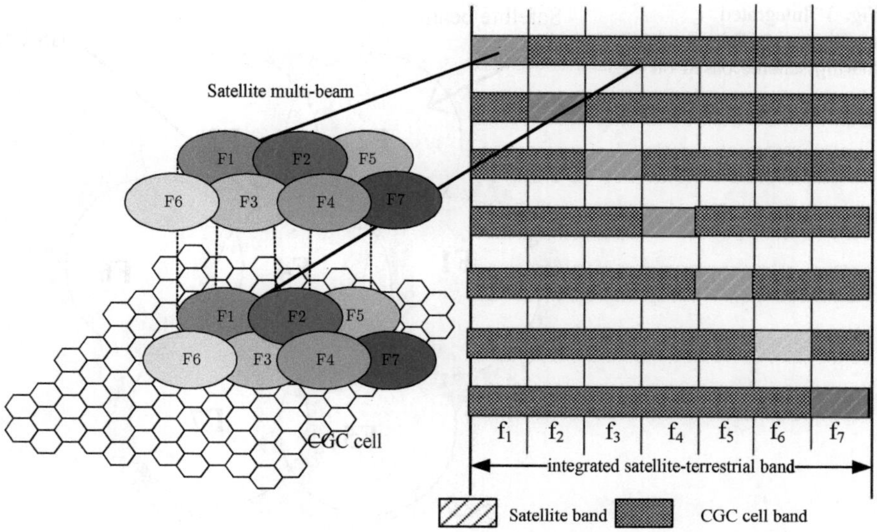

Fig. 2 Satellite-terrestrial spectrum 7-color multiplex diagram

system throughput. However, this scheme can cause a certain degree of co-frequency interference phenomenon between the satellite and terrestrial. Distance from the center of beam can lower the received satellite interference intensive. Therefore, the exclusion zone (EZ) is established around each satellite beam, the user or base station in EZ can not reuse frequency band used in the beam and only other frequency bands can be selected. The establishment of the exclusion zone can increase the angle to beam center, effectively reducing the antenna gain. Therefore, both the signal transmitted by satellite and the interference signal can be effectively reduced and satellite network user terminal or the terrestrial network user terminal signal to noise ratio and energy efficiency can be effectively improved.

3.2 Integrated Satellite-Terrestrial Spectrum Sharing Scheme Based on EZ

Firstly, the whole band in the integrated satellite-terrestrial network is divided into seven parts, seven bands are a cluster and the frequency reuse factor is 7. The base station in the peripheral area of the beam has serious interference to the satellite, and the satellite has serious interference to the terrestrial cell-edge user (CEU). Therefore, the establishment of EZ in the vicinity of beam area, users or base stations in the EZ can only select other frequency band. Based on the above principle, seven-color multiplexing of integrated satellite-terrestrial network are shown in Fig. 3.

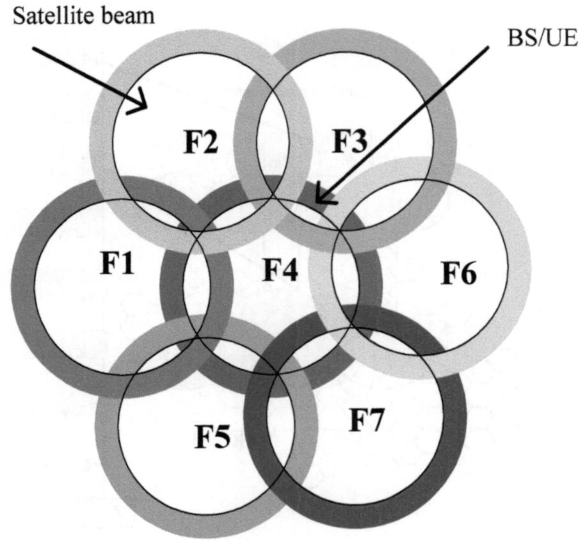

Fig. 3 Integrated satellite-terrestrial spectrum sharing scheme based on EZ

In case of seven-color multiplexing, the higher isolation frequency is set to a higher level of protection. Isolation degree from high to low in order to set the first-class protection, second-class protection, third-class protection and forbidden frequency band. The base station/UE prior uses the higher levels of protection frequency band to achieve lower co-frequency interference (Fig. 4).

To define the width of EZ by 3 dB (providing 3 dB isolation for the user or base station), take satellite to the terrestrial network terminal interference for example, the interference size calculation formula given by

$$P_{UE\_\mathrm{rec}} = P_{SAT} + G_{\mathrm{r}} + PL + P_{\mathrm{shadowloss}} \tag{14}$$

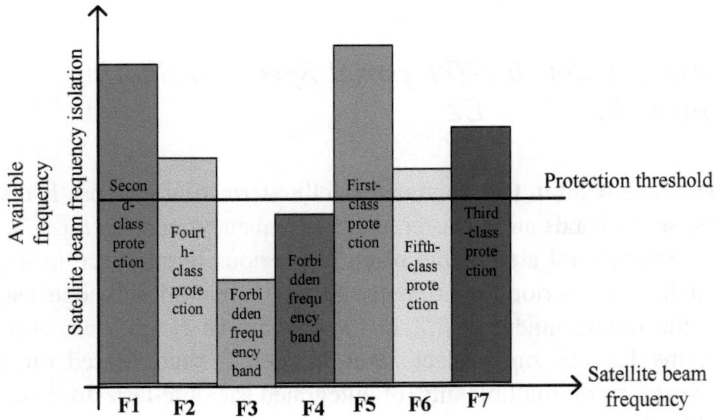

Fig. 4 Integrated satellite-terrestrial network frequency band isolation

where satellite antenna gain is represented by G_r and satellite-terrestrial link loss is represented by PL, where P_{SAT} and $PL + P_{shadowloss}$ basically remain the same. Therefore, reducing G_r can reduce the co-frequency interference.

3.3 Energy-Efficient Integrated Satellite-Terrestrial Spectrum Sharing Scheme

In the integrated satellite-terrestrial network, we propose an energy-efficient spectrum allocation scheme with high inter-cell fairness, combined with soft-frequency reuse strategy.

The terrestrial terminal located in the center of the terrestrial cell receive intensive useful signal and we can get high signal to noise ratio by using full frequency reuse scheme. At the same time, full-frequency reuse can ensure that integrated satellite-terrestrial network achieve the highest spectral efficiency and optimal throughput. In this chapter, an energy-efficient integrated satellite-terrestrial spectrum sharing scheme is proposed which assumes that the satellite beams are multiplexed with seven-color and the terrestrial cells are multiplexed with three-color soft frequencies. Energy-efficient integrated satellite-terrestrial spectrum sharing scheme is shown in Fig. 5.

In Fig. 5, cell 1, cell 2 and cell 3 are one cluster, located in the satellite beam F1 and the isolation between this cluster and surrounding six beams are shown in Fig. 5. In the plot area, the whole frequency multiplexing strategy is adopted in the red zone of the three central areas. In the traditional integrated satellite-terrestrial spectrum sharing scheme, the beam edge of cell 1 is multiplexed with F2, F3; the edge of cell 2 is multiplexed with F4, F5; the edges of cell 3 is multiplexed with F6 and F7. The energy-efficient integrated satellite-terrestrial spectrum sharing scheme based on fairness first ranks the satellite beam isolation of different frequency bands

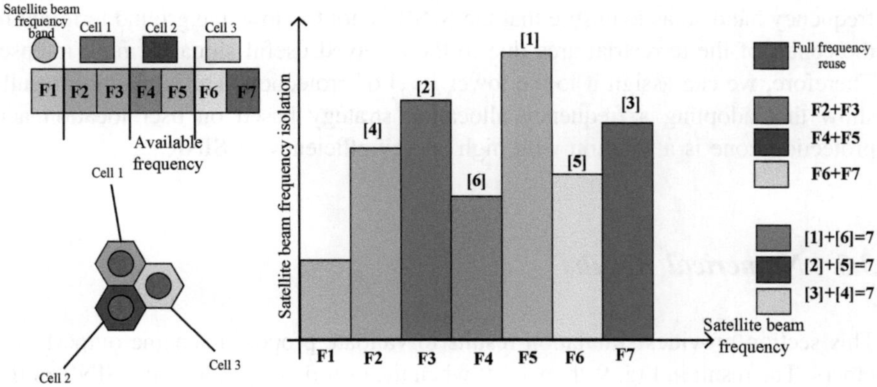

Fig. 5 Energy-efficient integrated satellite-terrestrial spectrum sharing scheme

corresponding to the base station/user location. The F4 with the highest degree of isolation and F5 with the lowest isolation of beam are divided into one group. F3 with sub-high degree of isolation and F6 with sub-low isolation of the beam are divided into one group and so on.

When the CEU access to the BS initially, we used higher isolation of frequency band. When user density is relatively small, we always use the better frequency band. However, when the high isolation frequency band can not effectively cover the users then we access the lower level isolation protection band. In this way, when the cell is not fully loaded, the energy efficient integrated satellite-terrestrial spectrum sharing scheme obviously has high SINR. When the cell tends to be full loaded, the energy efficient integrated satellite-terrestrial spectrum sharing scheme has good fairness among the cells.

3.4 *Joint Ue Location Efficient Energy Resource Management in Intergrated Satellite and Terrestrial Networks*

In addition, with the increase of protected areas, both downlink energy efficiency and SINR of terrestrial network terminals gradually increase, which is due to the fact that the size of the protected area directly reduces the size of the interference and the distance between the terrestrial terminal user and the satellite increases. The attenuation of the satellite signal link slightly increases, further reducing the interference caused by the satellite downlink. This is from the perspective of reducing interference to improve the system's SINR and energy efficiency. Based on the above simulation results, we propose a resource allocation scheme with higher energy efficiency. In the process of frequency allocation, we first consider establishing a protection zone around the satellite, and then consider whether the user of the ground network is located at the center or edge of the terrestrial cell. In the edge of the district we give priority to the allocation of higher protection frequency band so as to ensure that the SINR is not too low, the ground terminal in the center of the terrestrial area due to the received useful signal is more intense. Therefore, we can assign it to the lower level of protection. The simulation results show that adopting a frequency allocation strategy based on user location and protection zone is a solution with high energy efficiency of SINR.

3.5 *Numerical Results*

This section provides simulation results to validate proposed scheme outperforms others. The result in Fig. 9 shows that when the user density is low, the SINR of the cluster cell in new scheme is much better than others (Fig. 6).

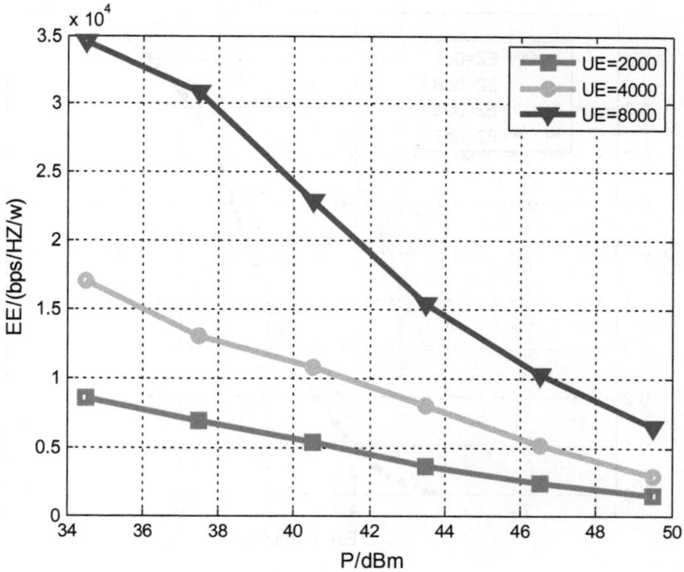

Fig. 6 Change of energy efficiency with terrestrial user transmitting power under different number of terrestrial satellite terminals

Simulation results show that the local surface network terminal number increased from 2000 to 8000, the energy efficiency of system presents the downward trend. The local surface network terminal number is huge, the terminal as the interference sources, the more the number, the terminal on satellite uplink interference is strong, satellite uplink terminal energy efficiency dropped significantly. In addition, as the launch power of the ground network terminal increases, the satellite receives the interference signal from the ground network terminal enhancement. Therefore, the energy efficiency of the satellite terminal uplink is decreasing.

Interference sources transmitted power size will affect the satellite link SINR, which influence the energy efficiency. Therefore, the more the number of ground terminals and the greater the terminal launch power, the lower the energy efficiency of the satellite link (Fig. 7).

The simulation results show that the bigger reserve satellite link is beneficial to the energy efficiency. Because as the reserve increase, accept satellite antenna gain is reduced, and the interference sources from satellite distance is also increased, the better anti-interference performance of the system (Fig. 8).

The closer the ground network terminal is to the base station, the stronger the received useful signal. The SINR and energy efficiency of the downlink of the terrestrial network terminal is higher, which is considered from increasing the received signal strength. In addition, with the increase of protected areas, both downlink energy efficiency and SINR of terrestrial network terminals gradually

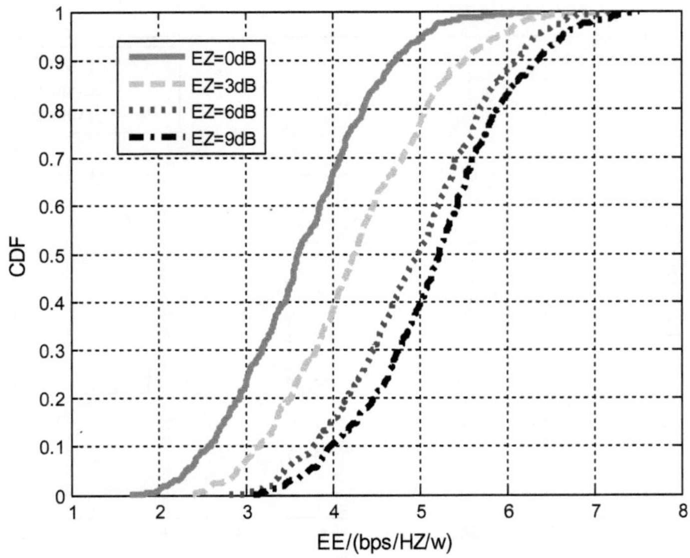

Fig. 7 The energy efficiency varies with the size of EZ

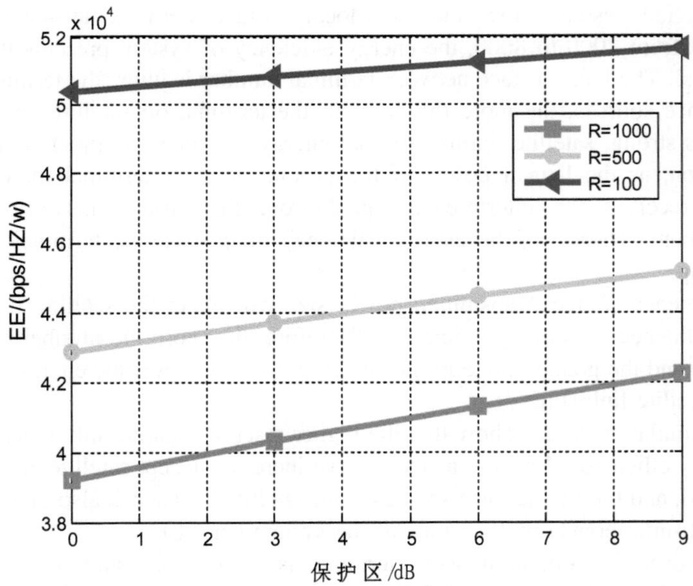

Fig. 8 Energy efficiency of terrestrial network terrestrial station at different positions from ground base station with the change of EZ

increase, which is due to the fact that the size of the protected area directly reduces the size of the interference and the distance between the terrestrial terminal user and the satellite increases, The attenuation of the satellite signal link slightly increases, further reducing the interference caused by the satellite downlink. This is from the perspective of reducing interference to improve the system's SINR and energy efficiency.

This section provides simulation results to validate proposed scheme outperforms others. The result in Fig. 9 shows that when the user density is low, the SINR of the cluster cell in energy efficient integrated satellite-terrestrial soft frequency reuse scheme is much better than the traditional satellite soft frequency reuse scheme.

Figure 10 shows that the SINR of the energy-efficient satellite sharing scheme proposed is 15 dB higher than the traditional scheme in cell 1 and 10 dB higher than the traditional scheme in cell 3. Therefore, when the user density is low, the proposed scheme has obvious advantages in SINR performance.

With the increasing number of users in the cell until the area is full load, the signal-to-noise ratio of energy efficient integrated satellite-terrestrial soft frequency reuse scheme declined. However, the SINR of the proposed scheme is superior to the traditional scheme as a whole.

Consider the comparison of the throughput of the two schemes under the worst case (three cells are full load). The simulation results are shown in Fig. 11.

Figure 11 shows that the throughput of energy-efficient satellite sharing scheme is close to each other in three cells when the cell is fully loaded. However, the traditional scheme of inter-cell throughput difference obviously and inter-cell user

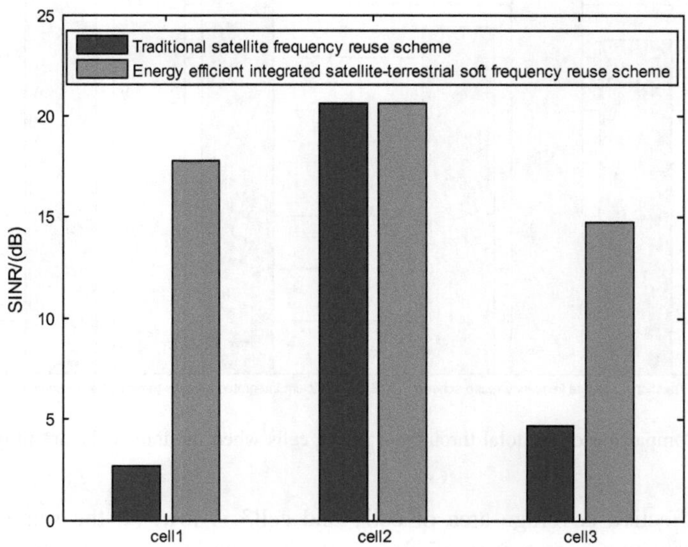

Fig. 9 The average SINR under the two schemes when the user density is low

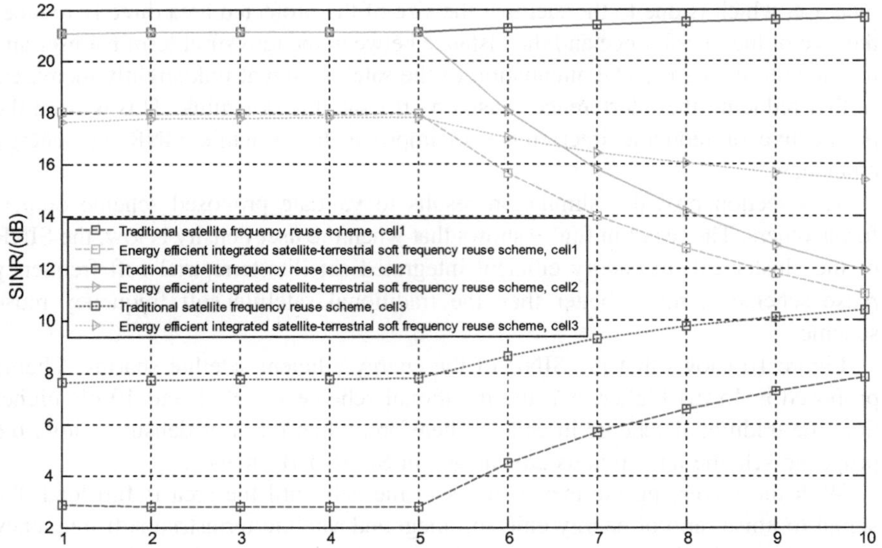

Fig. 10 The average SINR under the two schemes versus activity ratio of user

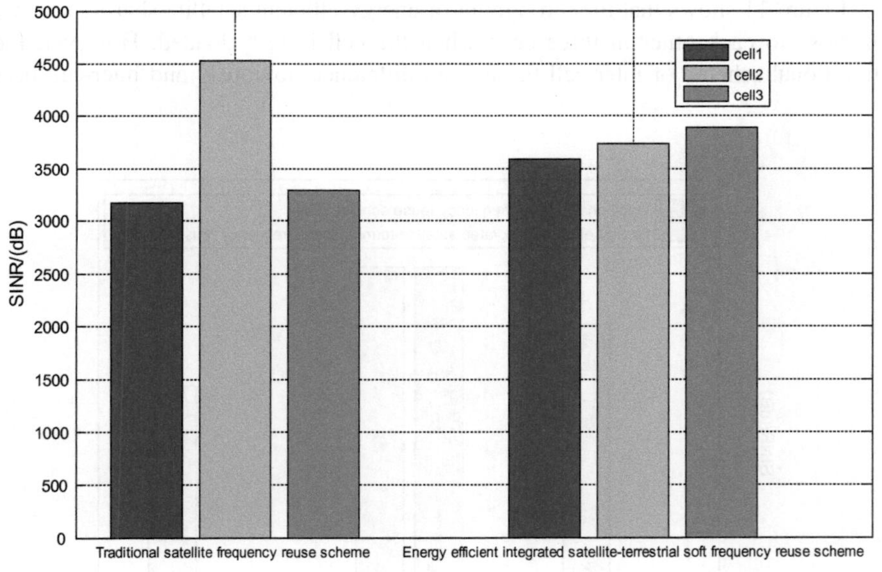

Fig. 11 Comparison of the total throughput of the cells when the three cells are fully loaded

unfair. Effective coverage area of cell1 and cell3 is low. At the same time, the total throughput of energy-efficient satellite sharing scheme in cluster is $3591 + 3738 + 3885 = 11214$ Mbit/s which is higher than that of the traditional scheme $3157 + 4536 + 3285 = 10978$ Mbit/s. Therefore, the energy-efficient satellite sharing scheme can take both the inter-cell fairness and throughput into account. Furthermore, simulate of the energy-efficiency at different user densities are shown in Fig. 12.

The simulation result of Fig. 12 shows that when the user density is small, the energy efficiency of the scheme proposed in this chapter has obvious advantages compared with the traditional scheme. As the user tends to be full load, the energy efficiency of the system is reduced, but the lowest value of the new scheme is still higher than the maximum value of the original scheme. This shows that the frequency allocation strategy proposed in this chapter is a kind of high energy efficiency scheme.

We have proposed energy-efficient integrated satellite-terrestrial spectrum sharing scheme based on inter-cell fairness. Firstly, the terrestrial terminal located in center of the terrestrial cell receive intensive useful signal and we can get high signal to noise ratio by using full frequency reuse scheme. Secondly, ranks the satellite beam isolation of different frequency bands corresponding to the base station/user location. Finally, we divide the highest degree of isolation band and the

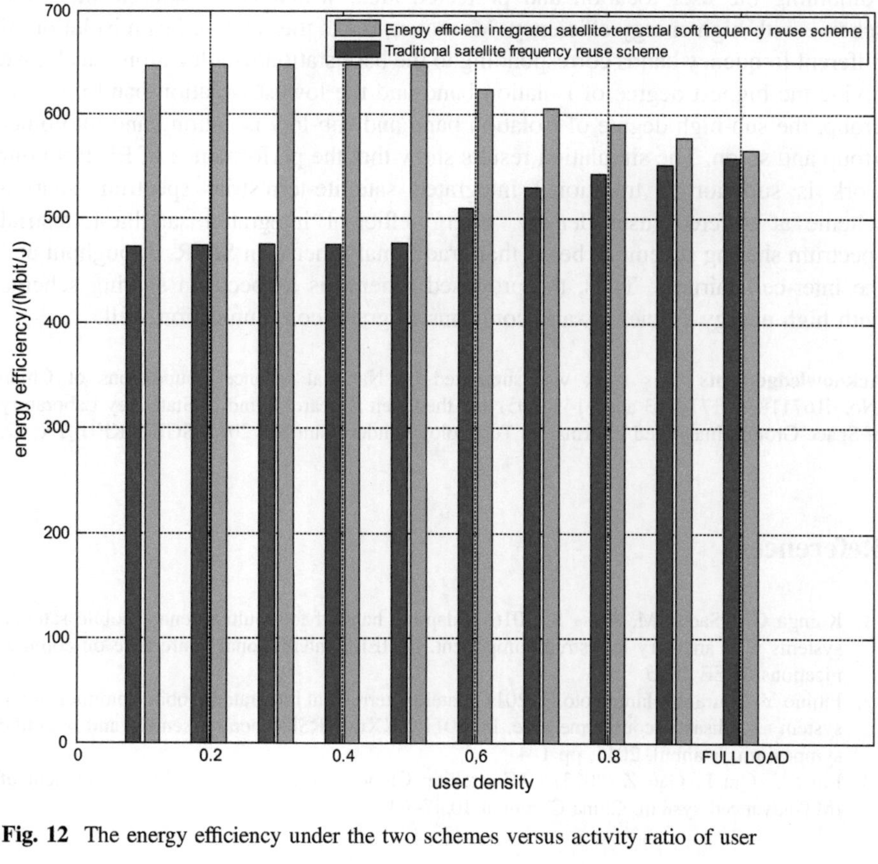

Fig. 12 The energy efficiency under the two schemes versus activity ratio of user

lowest isolation band into one group, the sub-high degree of isolation band and sub-low isolation band into other group and so on. The simulation results show that the performance of EE from our work is superior to traditional integrated satellite-terrestrial spectrum sharing scheme at different user density. Energy-efficient integrated satellite-terrestrial spectrum sharing scheme is better than traditional scheme in SINR, throughput and the inter-cell fairness.

4 Conclusion

In this chapter, we have proposed a satellite–terrestrial energy-efficient spectrum-sharing scheme based on UE location. We first classify the users in the satellite–earth integrated network into two categories (satellite network terminal users and terrestrial network users) and establish the satellite–earth integrated network architecture. Secondly, the terrestrial terminal located in the center of the terrestrial cell receives an intense useful signal, and we can obtain a high SINR by using a full frequency-reuse scheme. We proposed a spectrum-sharing scheme combining the user location and protected area, which is a resource allocation scheme with high energy efficiency. Moreover, ranks the satellite beam isolation of different frequency bands corresponding to the base station/user location. Lastly, we divide the highest degree of isolation band and the lowest isolation band into one group, the sub-high degree of isolation band and sub-low isolation band into other group and so on. The simulation results show that the performance of EE from our work is superior to traditional integrated satellite-terrestrial spectrum sharing scheme at different user density. Energy-efficient integrated satellite-terrestrial spectrum sharing scheme is better than traditional scheme in SINR, throughput and the inter-cell fairness. Thus, the proposed scheme is a spectrum sharing scheme with high energy efficiency, and conforms to green communication well.

Acknowledgements This work was supported by National Science Foundations of China (No. 61671183, 61771163 and 91438205) and the Open Research Fund of State Key Laboratory of Space-Ground Integrated Information Technology under grant No. 2015_SGIIT_KFJJ_TX_02.

References

1. Kamga GN, Sadek M, Aïssa S (2016) Adaptive handoff for multi-antenna mobile satellite systems with ancillary terrestrial component. In: IEEE international conference on communications, IEEE 2016
2. Fujino Y, Miura A, Hamamoto N (2011) Satellite terrestrial integrated mobile communication system as a disaster countermeasure. In: 2011 XXXth URSI general assembly and scientific symposium, Istanbul, 2011, pp 1–4
3. Liu SY, Qin F, Gao Z (2013) LTE-satellite: Chinese proposal for satellite component of IMT-advanced system. China Commun 10:47–64

4. Orsino A, Araniti G, Scopelliti P et al (2017) Optimal subgroup configuration for multicast services over 5G-satellite systems. In IEEE international symposium on broadband multimedia systems and broadcasting, IEEE 2017
5. Zhu X, Jiang C, Kuang L et al (2017) Non-orthogonal multiple access based integrated terrestrial-satellite networks. IEEE J Sel Areas in Commun (99):1–1
6. Deslandes V, Tronc J, Beylot AL (2010) Analysis of interference issues in integrated satellite and terrestrial mobile systems. In: Advanced satellite multimedia systems conference, 2010, pp 256–261
7. Giambene G, Le VA, Bourgeau T et al (2015) Soft frequency reuse schemes for heterogeneous LTE systems In: IEEE international conference on communications, IEEE 2015, pp 3161–3166
8. Xie H, Wang B, Gao F et al (2016) A full-space spectrum-sharing strategy for massive MIMO cognitive radio system. IEEE J Sel Areas Commun 34(10):2537–2549
9. Park U, Kim HW, Oh DS et al (2016) Performance analysis of dynamic resource allocation for interference mitigation in integrated satellite and terrestrial systems In: International conference on next generation mobile applications, services and technologies, IEEE 2016, pp 217–221
10. Sheng M, Wang Y, Li J, Liu R, Zhou D, He L (2017) Toward a flexible and reconfigurable broadband satellite network: Resource management architecture and strategies. IEEE Wireless Commun 24(4):127–133, Aug 2017. http://doi.org/10.1109/MWC.2017.1600173
11. Jia M, Gu X, Guo Q et al (2016) Broadband hybrid satellite-terrestrial communication systems based on cognitive radio toward 5G. IEEE Wirel Commun 23(6):96–106
12. Vannithamby R, Talwar S, Chih-Lin I et al (2016) Towards green and soft towards 5G. Wiley, 53–77
13. Sheng M, Wang Y, Li J et al (2017) Toward a flexible and reconfigurable broadband satellite network: resource management architecture and strategies. IEEE Wirel Commun 99:2–8
14. Min Jia, Zhang Ximu Gu, Xuemai Liu Xiaofeng, Qing Guo (2018) Joint UE location energy-efficient resource management in integrated satellite and terrestrial networks. J Commun Inf Netw 3(1):61–67 SD
15. GPP TR 36.942 V10.3.0 (2012) Technical specification group radio access network, evolved universal terrestrial radio access (E-UTRA), radio frequency (RF) system scenarios(release 10), 2012, pp 14–15
16. Jia M, Zhang X, Gu X, Guo Q (2018) Energy efficient cognitive spectrum sharing scheme based on inter-cell fairness for integrated satellite-terrestrial communication systems. In: 2018 IEEE VTC Spring. 3–6 June 2018, Porto, Portugal
17. Jia M, Gu X, Guo Q, Xiang W, Zhang N (2016) Broadband hybrid satellite-terrestrial communication systems based on cognitive radio toward 5G. IEEE Wirel Commun 23(6): 96–106

Intelligent Sub-meter Localization Based on OFDM Modulation Signal

Mu Zhou, Ze Li, Yue Jin, Zhenyuan Zhang and Zengshan Tian

Abstract With the development of Internet of Thing (IoT), high accuracy positioning is of a significant importance in indoor environment. In recent years, localization technology has advanced greatly aiming at localizing the target by utilizing existing infrastructure such as LTE and Wi-Fi. However, indoor localization still faces many challenges caused by the complex indoor environment. In this chapter, first of all, a new indoor localization approach by employing the Angle-of-arrival (AOA) and Received Signal Strength (RSS) measurements in Line-of-Sight (LOS) environment is proposed. Specifically, the Channel State Information (CSI) is collected by using the commodity Wi-Fi devices with our designed three antennas to estimate the AOA of Wi-Fi signal. Second, a new direct path identification algorithm is proposed to obtain direct signal path for the sake of reducing the interference of multipath effect on AOA estimation. Third, a new objective function is constructed to solve localization problem by integrating AOA and RSS information. Although the localization problem is non-convex, the Second-order Cone Programming (SOCP) relaxation approach is used to transform it into a convex problem. Finally, the effectiveness of the proposed approach is verified based on the prototype implementation by using the commodity Wi-Fi devices. The experimental results show that this approach can achieve the median error 0.7 m in the actual LOS indoor environment. However; the localization in

M. Zhou (✉) · Z. Li · Y. Jin · Z. Zhang · Z. Tian
Chongqing Key Lab of Mobile Communications Technology,
Chongqing University of Posts and Telecommunications,
Chongqing 400065, China
e-mail: zhoumu@cqupt.edu.cn

Z. Li
e-mail: 574625020@qq.com

Y. Jin
e-mail: 702814793@qq.com

Z. Zhang
e-mail: zhangzhenyuan163@163.com

Z. Tian
e-mail: tianzs@cqupt.edu.cn

T. S. Durrani et al. (eds.), *Geological Disaster Monitoring Based on Sensor Networks*, Springer Natural Hazards, https://doi.org/10.1007/978-981-13-0992-2_11

Non-line-of-sight (NLOS) environment is challenging since the incorrect AOA introduces large localization errors. Therefore, sensing the LOS or NLOS path between the AP and target is significantly important for improving the performance of a large number of mobile computing applications such as indoor localization. Various promising systems in current commodity Wi-Fi network rely on the frequency-dependent amplitude and phase of CSI to construct the LOS and NLOS identification scheme, but the corresponding robustness is not considered enough. With the development of Multiple-input-multiple-output (MIMO) technology with smart antennas, the spatial property of channel can be obtained effortlessly. Consequently, a potential is provided to utilize the unchanged spatial information of multipath signals to deliver robustness identification system. In this chapter, the Wi-vision, an accurate and robust LOS identification system based on indoor Single-input-multiple-output (SIMO) measurements, is proposed. The concept of Hopkins statistic is introduced to measure the distribution properties of the AOA and relative Time-of-flight (TOF) under the LOS and NLOS conditions. The experimental results show that the Wi-vision is able to achieve the LOS and NLOS detection rates above 91 and 87% respectively under different system configuration.

Keywords Indoor localization · OFDM · WiFi · Channel state information

1 Introduction

With the continuous development and progress of science and technology, the demand for wireless communication is constantly changing and improving. Therefore, in improving and developing wireless communication technology, the existing available frequency band is also greatly reduced. How to make better use of the undeveloped frequency band and make the spectrum utilization more effective has become an important problem to be solved urgently. In many wireless mobile communication technologies, OFDM (Orthogonal Frequency Division Multiplexing) is one of the most promising technologies. Orthogonal frequency division multiplexing technology is a high utilization rate of frequency spectrum modulation technique was proposed in twentieth Century 50 to 60s, the basic idea is through the spectrum overlap of sub channels is allowed, and the mutual influence of frequency division multiplexing method to parallel data transmission OFDM system has not only the spectrum utilization rate high, and have strong ability of anti-multipath. Due to technical limitations and shortcomings, OFDM technology has not been widely concerned and applied when it has just been proposed. Until the 70s of last century, the discrete Fourier inverse transformation method was used to realize the multicarrier modulation more simply. On this basis, it laid a solid foundation for the application of OFDM technology in the actual communication system. More and more scholars have launched a lot of research on OFDM system and related technologies, and these researches have greatly accelerated the development process of OFDM in the field of actual wireless communication. At present,

OFDM technology has been widely used in digital video broadcasting wirelesses LAN, high definition TV and so on, and its application scene is becoming more and more common.

2 Principle

In the frequency domain, multiple adjacent tones or subcarriers are independently modulated with complex data. As shown in Fig. 1, an IFFT (Inverse Fast Fourier Transform) is performed on the frequency-domain subcarriers to produce the OFDM symbol in the time-domain. Then, in the time domain, guard intervals are inserted between the symbols to prevent inter-symbol interference at the receiver caused by multi-path delay spread in the radio channel. Multiple symbols can be concatenated to create the final OFDM burst signal. At the receiver, a FFT (Fast Fourier Transform) is performed on the OFDM symbols to recover the original data bits.

The OFDM modulator includes several modules: training sequence generation module, scrambling module, channel encoding module, interleaving module, modulation module, pilot insertion module, IFFT module and cyclic prefix module. The input data for the serial data by scrambling, encoding, interleaving and QAM modulation mapping become parallel data; these data will be parallel in the insertion of the pilot symbols after IFFT operation, operation after adding cyclic prefix, and the last training sequences into a complete data for output. OFDM demodulator including detecting synchronization module, carrier synchronization module, symbol synchronization module, frequency synchronization module, demodulation module. The received data must first determine whether there is synchronous detection, the new data coming; then the data carrier synchronization, ensure the orthogonally of the subcarriers: after symbol synchronization, then the exact location of the data; the last one is the synchronous sampling frequency synchronization, sampling frequency of the transmitter and receiver side compensation; because it is in the frequency domain of the data sampling frequency offset

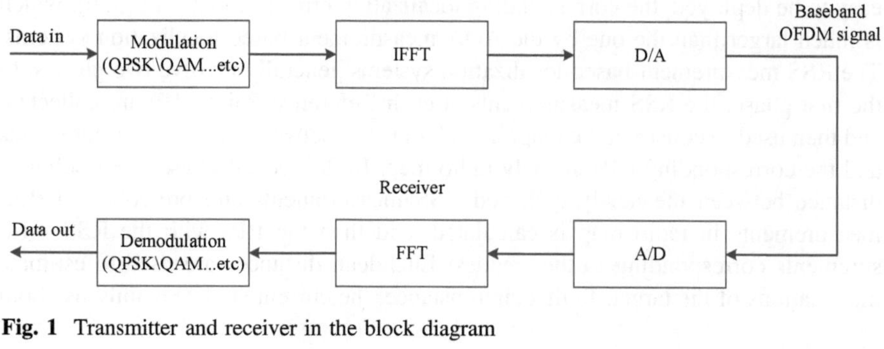

Fig. 1 Transmitter and receiver in the block diagram

compensation, so the need for sampling frequency synchronization before FFT operation: after the completion of synchronization needed after QAM demodulation to get the correct data.

3 OFDM Based Localization Systems

3.1 Introduction

With the increase of the demand for the ubiquitous Location-based Services (LBSs), localization and navigation applications become more important in daily life. In an outdoor environment, people prefer to use the Global Navigation Satellite System (GNSS) to achieve the LBSs, whereas the signal from the satellites cannot be easily received in indoor environment, which results in the failure of the GNSS. However, many LBSs significantly depend on the highly-accurate indoor localization such as when finding goods in the mall, locating in the mine, and rescuing.

Wi-Fi localization has great promise in the area of indoor localization due to the wide deployment of the commodity Wi-Fi devices. Up to now, existing Wi-Fi localization systems are mainly based on the Angle-of-arrival (AOA) or Received Signal Strength (RSS) measurement. The accuracy of the AOA measurement-based localization systems is around 0.4 m [1, 2]. They measure the AOA of the signal from at least two Access Points (APs), and then use the triangulation algorithm to locate the target. By using the well-known Multiple Signal Classification (MUSIC) algorithm [3], each AP is equipped with six antennas for the AOA estimation since there are normally about six significant signal paths in indoor environment [4]. Thus, the AOA measurement-based localization systems depend on the special hardware modification like the design of specific antennas, which is challenging and with significantly high cost. To solve this problem, we propose a new approach by using the Channel State Information (CSI) which is available in many existing commodity Wi-Fi devices to estimate the AOA of the multipath signal with small hardware modification.

At the same time, the RSS measurement-based localization systems have also become popular by using the existing commodity Wi-Fi devices. Although they are easy to be deployed, the corresponding localization error is about 3 m [5, 6], which is much larger than the one by the AOA measurement-based localization systems. The RSS measurement-based localization systems generally involve two phases. In the first phase, the RSS measurements at each Reference Point (RP) are collected, and then used to construct the mapping relationship between the RSS measurements and the corresponding RPs, namely radio map. In the second phase, the Euclidean distance between the newly collected RSS measurements and pre-collected RSS measurements in radio map is calculated, and then the RPs with the RSS measurements corresponding to the smallest Euclidean distances are used to estimate the locations of the target. In this circumstance, the current studies mainly use both the AOA and RSS measurements to construct anew objective function for the indoor localization problem. On one hand, this approach improves the localization accuracy compared with the one by using the RSS or AOA measurement solely. On the other hand, it significantly reduces the time and labor cost involved in the radio map construction, and also exhibits well robustness to the environmental change.

In summary, the main contribution of this approach is to design a new precise indoor localization system based on the CSI from the commodity Wi-Fi APs to estimate the AOA of the multipath signal. To overcome the limitation of the conventional AOA estimation approaches, which generally require at least six antennas, it relies on the OFDM modulation property to estimate the AOA of the signal with only three antennas. This approach is based on the fact that the multipath effect not only results in the measurable change of the CSI in the antenna array, but also changes the CSI across different sub-carriers due to the difference of the Time of Arrival (TOA) measurement. Specifically, when the signal with the AOA arrives at the linear antenna array which consists of *M* antennas, the difference of the propagation delay between two neighboring antennas with the spacing distance *d* is $d \times \sin\theta$, as shown in Fig. 2. In addition, since the Wi-Fi signal is based on the OFDM modulation and the different subcarriers are with different frequencies, the different subcarriers will be featured with different accumulative phases with respect to the same TOA measurement.

In the indoor environment, the Wi-Fi signal from the same AP is generally correlated, which is not appropriate for the conventional MUSIC algorithm. To solve this problem, the current studies propose to use the two-dimensional spatial smoothing approach for the AOA estimation with respect to the multiple correlated signals. Furthermore, the CSI measured by the Wi-Fi Network Interface Card (NIC) suffers from the measurement error due to the imperfect signal processing by the hardware like the signal boundary detection, which introduces the additional time delay, namely Packet Detection Delay (PDD), to all the signal paths. Since different signal packets are with different PDD, the TOA measurement cannot capture the true time taken by the signal traveling from the APs to the target. In this circumstance, the current studies use the CSI from multiple APs to estimate the AOA and TOA measurements of the multipath signal to locate the target. However, the most important problem is the identification of the direct signal path from the target to each AP. The authors in [7] proposed to identify the direct signal path

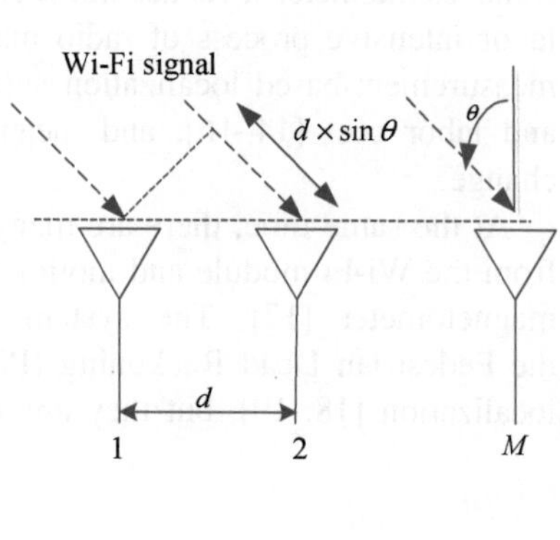

Fig. 2 Wi-Fi signal arriving at linear antenna array

according to the TOA measurement, which may not be reliable in the multipath environment. The authors in [8] declared that the direct signal path is corresponding to the AOA measurement with the highest space spectrum value. However, in the indoor environment, the direct signal path may be weaker than the indirect path resulted from the multi-path effect.

Based on the assumption that the signal on the reflected paths is generally with large variation of the AOA and TOA measurements compared with the one on the direct path, the Gaussian means clustering algorithm can be utilized to identify the direct signal path. By considering the fact that the error of the CSI results in the additional noise in the TOA estimation, the phase sanitization algorithm can be used to remove the interference of this error before estimating the AOA and TOA measurements.

Different from the conventional indoor localization systems which integrate multiple types of measurements [9, 10], the current studies propose to use a new linear Least Square (LS)-based object function for the localization. In addition, since the localization problem is non-convex, they use the Second-order Cone Programming (SOCP) relaxation approach to convert it into a convex problem. The findings are summarized as follows. First of all, the proposed system is easy to be deployed and with small hardware modification. Second, by using the commodity Wi-Fi APs with three antennas, the receiver is independent of the motion sensors like the gyroscope and accelerometer. Third, the proposed system achieves the median error 0.7 m and 68th error 1 m, which are smaller than the ones by the conventional systems using the AOA or RSS measurement solely.

3.2 Related Work

The existing Wi-Fi localization systems are mainly based on the propagation modeling [11] and location fingerprinting [12]. The systems using the propagation modeling locate the target based on the triangulation approach, while the ones using the location fingerprinting locate the target by constructing the mapping relationship between the RSS patterns and physical locations. One of the most representative location fingerprinting-based localization systems is the RADAR [13], which achieves the meter-level accuracy. However, it suffers from the time consuming and labor intensive process of radio map construction. There are some evolved RSS measurement-based localization systems which are independent of the huge time and labor cost [14–16], and mean while they are robust to the environmental change.

At the same time, there are many fusion localization systems by using the data from the Wi-Fi module and motion sensors like the gyroscope, accelerometer, and magnetometer [17]. The systems using the motion sensors generally apply the Pedestrian Dead Reckoning (PDR) algorithm to achieve the continuous-time localization [18, 19], but they are suffered by the accumulative error as the time goes. However, the fusion localization systems are limited for the application since many mobile devices are not embedded with motion sensors.

Since many off-the-shelf Wi-Fi APs supports the Multiple Input Multiple Output (MIMO) technique by using multiple antennas, the AOA measurement-based localization systems have been significantly developed. The Array Track [20] requires the Wi-Fi APs to be equipped with at least seven antennas to estimate the AOA measurement. The Ubicarse [1] is based on the Synthetic Aperture Radar (SAR) technique to achieve the sub-meter localization accuracy. However, it requires the target to be equipped with a rotation antenna which cannot be satisfied by the commodity Wi-Fi devices. Recently, there are some systems using the commodity Wi-Fi devices to estimate AOA measurement [21].

3.3 System Descriptions

As shown in Fig. 3, the proposed system consists of three main steps as follows.

1. **CSI-based AOA estimation**. This system estimates the AOA and TOA based on the CSI obtained from the existing commodity Wi-Fi APs with three antennas, and meanwhile employs the 2-D spatial smoothing algorithm to eliminate the interference of the coherent multi-path signal.
2. **Direct signal path identification**. This system identifies the direct signal path based on the likelihood of each cluster obtained by the Gaussian means clustering algorithm, and meanwhile uses the phase sanitization algorithm to avoid the impact of the error of CSI measurement on the TOA estimation.
3. **Target localization**. This system integrates the AOA and RSS measurements based on the LS criterion to construct a new object function for the localization problem.

In this system, the transmitter is required to conform to the IEEE 802.11n standard for the sake of employing the spatial diversity technique with multiple antennas to achieve the high data transmission rate. Figure 4 shows the flow chart of signal processing with the 802.11n NIC. The incoming analog signal, $s(t)$, is processed by the Automatic Gain Controller (AGC) to compensate the signal amplitude attenuation, and then sampled as the discrete signal, $s(n)$. The packet

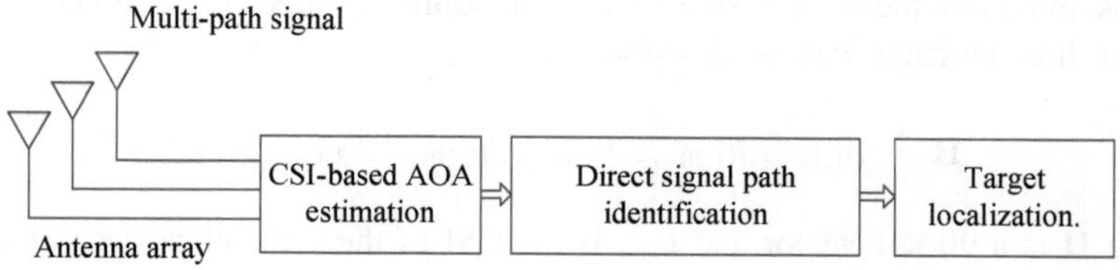

Fig. 3 System description

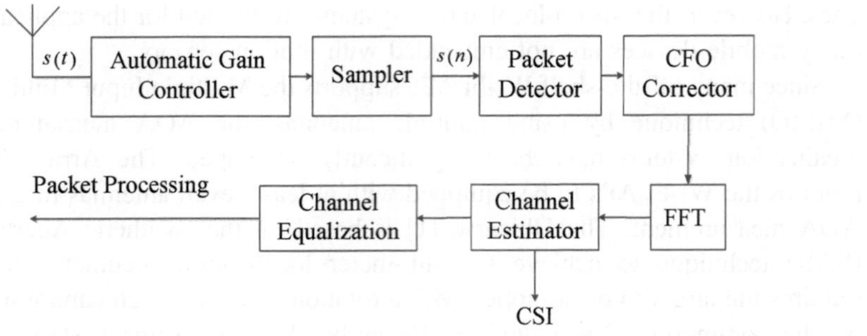

Fig. 4 Flow chart of signal processing with the 802.11n NIC

detector and Central Frequency Offset (CFO) corrector are used to confirm the incoming packet and compensate the central frequency offset respectively. To extract the data correctly, the receiver applies the channel equalization to estimate the impact of the channel on each subcarrier [22]. The CSI obtained from the channel equalization involves the amplitude and phase information.

In the indoor environment, there are always many signal paths detected at the receiver due to various signal refraction and reflection. Different signal paths are featured with different attenuation and propagation delay, while the corresponding RSS measurements are generally assumed to obey the Gaussian distribution [23]. Based on the results in [24], the Channel Frequency Response (CFR) can be described as

$$h(f) = \sum_{l=0}^{L} \gamma_l \cdot e^{-j \cdot 2\pi \cdot f \cdot \tau_l} \tag{1}$$

where L is the number of signal paths. γ_l and τ_l are the path coefficient and propagation delay respectively with respect to the l-th signal path. The CSI is obtained by sampling the CFR with the sampling rate $F_f = 1/\Delta f$ and Δf is the sampling interval. In this system, the Intel 5300 NIC is selected to report the CSI of 30 subcarriers at each antenna.

3.3.1 Two-dimensional Spatial Smoothing

By assuming that there are K signal paths including the direct and indirect ones, the CSI at three antennas can be described as

$$\mathbf{H} = \left[h_{1,1}, \ldots h_{1,30}, \ldots h_{2,1}, \ldots h_{2,30}, \ldots h_{3,1}, \ldots h_{3,30}\right]^T \tag{2}$$

where $\mathbf{H}$ is a 90×1 vector and $h_{m,n}$ is the CSI of the n-th subcarrier at the m-th antenna. Based on Eqs. (1) and (2), we obtain

$$\mathbf{H} = \mathbf{A}\boldsymbol{\Gamma} + \mathbf{N} \tag{3}$$

where $\boldsymbol{\Gamma} = [\gamma_1 \cdots \gamma_K]$ is a $K \times 1$ vector of the path coefficient with respect to the K signal paths and $\mathbf{N}$ is a 90×1 noise vector. $\mathbf{A}$ is the $90 \times K$ direction matrix and $\mathbf{A}$ equals to

$$\mathbf{A} = [\mathbf{a}(\theta_1, \tau_1), \ldots, \mathbf{a}(\theta_k, \tau_k), \ldots, \mathbf{a}(\theta_K, \tau_K)]^{\mathrm{T}} \tag{4}$$

where $\mathbf{a}(\theta_k, \tau_k)$ is a 90×1 direction vector which is described as

$$\mathbf{a}(\theta_k, \tau_k) = [\mathbf{a}_1(\theta_k, \tau_k), \mathbf{a}_2(\theta_k, \tau_k), \mathbf{a}_3(\theta_k, \tau_k)]^T \tag{5}$$

In Eq. (5), $\mathbf{a}_m(\theta_k, \tau_k)$ is a 30×1 direction vector of the k-th signal path at the m-th antenna, as shown in Eq. (6).

$$\mathbf{a}_m(\theta_k, \tau_k) = \left[a_{1,m}(\theta_k, \tau_k), \ldots, a_{i,m}(\theta_k, \tau_k), \ldots, a_{30,m}(\theta_k, \tau_k)\right]^T \tag{6}$$

where $a_{i,m}(\theta_k, \tau_k) = e^{-j\Delta\psi_{i,m}(k)}$ and $\Delta\psi_{i,m}(k) = 2\pi(i-1)\Delta f \tau_k + 2\pi d(m-1)\frac{\sin\theta_k}{\lambda_i}$ is the phase difference of the i-th ($i = 1, \ldots, 30$) subcarrier. Δf is the frequency interval of every two neighboring subcarriers. d is the physical distance between every two adjacent antennas. θ_k and τ_k are the AOA and TOA with respect to the k-th signal path respectively.

The covariance matrix of the CSI in Eq. (3) is calculated by

$$\mathbf{R}_{\mathrm{x}} = E\{\mathbf{H}\mathbf{H}^{\mathrm{H}}\} \tag{7}$$

where $\mathbf{H}^{\mathrm{H}}$ is the conjugate transpose of $\mathbf{H}$. Based on the result in [25], the eigenvectors corresponding to the noise are orthogonal to the direction vectors in $\mathbf{A}$ and the space spectrum of the AOA and TOA, θ and τ, can be described as

$$\mathrm{P}_{music}(\theta, \tau) = \frac{1}{\mathbf{a}^{\mathrm{H}}(\theta, \tau)\mathbf{E}_{\mathrm{N}}\mathbf{E}_{\mathrm{N}}^{\mathrm{H}}\mathbf{a}^{\mathrm{H}}(\theta, \tau)} \tag{8}$$

where $\mathbf{E}_{\mathrm{N}}$ is the set of eigenvectors with respect to the noise subspace of $\mathbf{R}_{\mathrm{X}}$. To eliminate the interference of the coherent signals, this system conducts the Two-dimensional Spatial Smoothing (2D-SS) [26], on $\mathbf{R}_{\mathrm{X}}$ instead of $\mathbf{H}$, as shown in Fig. 5. The elements in the dashed green and red boxes construct the covariance matrices of the first and second sub-arrays respectively. Based on the observation that the first elements of the covariance matrices in the first and second sub-arrays are $h_{1,1} \times h_{1,1}$ and $h_{1,2} \times h_{1,2}$, this system obtains the covariance matrices of the existing sub-arrays by increasing the subcarrier ID and antenna index number to $L_2 = 30 - N_{sub2} + 1$ and $L_1 = 3 - N_{sub1} + 1$ respectively, where $N_{sub1} = 2$ and $N_{sub2} = 15$. Then, the number of sub-arrays and elements in each sub-array equals to $L = L_1 \times L_2$ and $N_{sub1} \times N_{sub2}$ respectively.

Fig. 5 Process of 2D-SS on $\mathbf{R}_X$

$$\begin{bmatrix} h_{1,1}\times h_{1,1} & \cdots & h_{1,1}\times h_{1,30} & h_{1,1}\times h_{2,1} & \cdots & h_{1,1}\times h_{3,30} \\ \vdots & h_{1,2}\times h_{1,2} & \vdots & \vdots & & \vdots \\ h_{1,30}\times h_{1,1} & \cdots & h_{1,30}\times h_{1,30} & h_{1,30}\times h_{2,1} & \cdots & h_{1,30}\times h_{3,30} \\ h_{2,1}\times h_{1,1} & \cdots & h_{2,1}\times h_{1,30} & h_{2,1}\times h_{2,1} & \cdots & h_{2,1}\times h_{3,30} \\ \vdots & & \vdots & \vdots & & \vdots \\ h_{3,30}\times h_{1,1} & \cdots & h_{3,30}\times h_{1,30} & h_{3,30}\times h_{2,1} & \cdots & h_{3,30}\times h_{3,30} \end{bmatrix}$$

The covariance matrix of the CSI after the process of 2D-SS on $\mathbf{R}_X$, $\mathbf{R}_{2D-SS}$, is modified into

$$\mathbf{R}_{2D-SS} = \frac{1}{L_1 L_2}\sum_{m=1}^{L_1}\sum_{n=1}^{L_2}\mathbf{R}_{m,n} \tag{9}$$

where $\mathbf{R}_{m,n}$ is the sub-covariance matrix in $\mathbf{R}_X$ with respect to the n-th subcarrier at the m-th antenna. Then, we conduct the MUSIC algorithm on the smoothed covariance matrix of the CSI to obtain the direction vectors, as well as the AOA and TOA with respect to each signal path.

To verify the effectiveness of the 2D-SS in AOA estimation, two simulations are conducted as follows. This system assumes that there are 30 subcarriers operating at the 5.2 GHz with the spacing distance 1.25 MHz. The antenna array contains three antennas and the spacing distance between every two adjacent antennas, d, equals to $\lambda/2$, where λ is the wavelength of the signal. The impact of value d on the AOA estimation will be further discussed in the following.

In the first simulation, it is assumed that there are four incoherent signals with the AOA-20°, −10°, 10°, 20° and TOA 10, 30, 20, and 60 ns respectively. Figure 6 shows the results of the AOA and TOA estimation, from which it can be found that by the 2D-SS, the estimated AOA and TOA are much similar to the real ones.

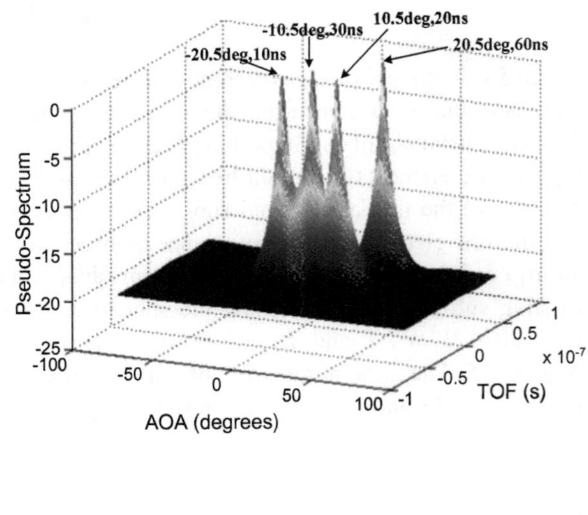

Fig. 6 Results of AOA and TOA estimation for the incoherent signals

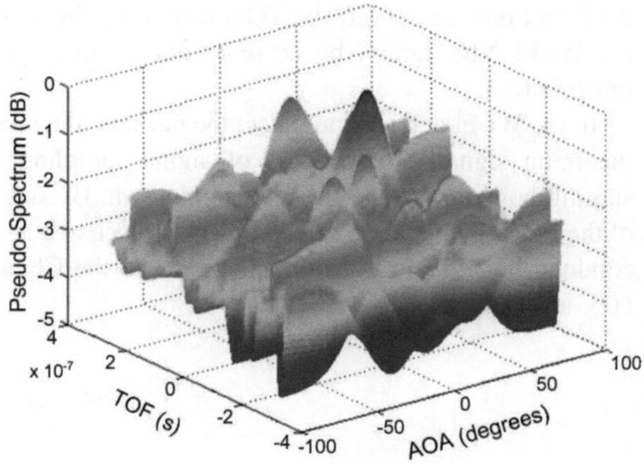

Fig. 7 Results of AOA and TOA estimation for the coherent signals without the 2D-SS

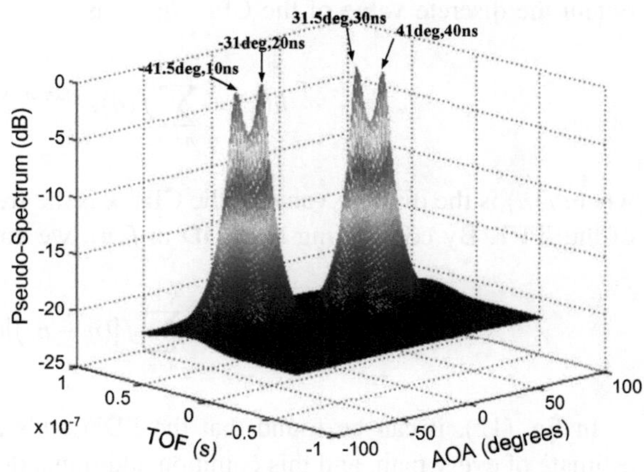

Fig. 8 Results of AOA and TOA estimation for the coherent signals with the 2D-SS

In the second simulation, it is assumed that there are four coherent signals with the AOA −40°, −35°, 35°, 40° and TOA 10, 20, 30, and 40 respectively. Figures 7 and 8 show the results of the AOA and TOA estimation without and with the 2D-SS.

From these figures, it is concluded that the 2D-SS can significantly improve the accuracy of the AOA and TOA estimation for both the incoherent and coherent signals.

3.3.2 Direct Signal Path Identification

Packet Detection Delay

To identify the direct signal path, many prior studies [27, 28] rely on the TOA measurement to claim that the signal path with the shortest TOA is most likely to be the direct one. However, the TOA estimation by using the raw CSI obtained from the Wi-Fi NIC cannot be accurate due to the channel distortion and hardware imperfection.

In the Wi-Fi network, to detect the packets, the receiver is required to sample the incoming signal. The process of signal sampling involves the PDD since the starting boundary of the packets is unknown. By setting n_ξ as the PDD, the impact of the PDD on the phase measurement is discussed as follows. Based on Eq. (1), we conduct the IFFT to transform the CFR to the Channel Impulse Response (CIR), $f(t)$, as follows.

$$f(t) = \sum_{l=0}^{L} \gamma_l \delta(t - \tau_l) \tag{10}$$

where $\delta(\cdot)$ is the delta function. Then, the discrete Fourier transform is conducted to obtain the discrete value of the CFR, $h(k)$, as

$$h(k) = \sum_{n=0}^{N-1} f(n) e^{-j2\pi kn/N} \tag{11}$$

where $f(n)$ is the discrete value of the CIR. k is the frequency index. N is the length of the IFFT. By considering the PDD in $f(n)$, we convert Eq. (11) into

$$h(k) e^{-j2\pi k n_\xi/N} = \sum_{n=0}^{N-1} f[(n - n_\xi)] e^{-j2\pi kn/N} \tag{12}$$

In Eq. (12), it can be found that the PDD adds a constant offset to the TOA estimate of every path, and this common additional delay manifests itself as a linear in frequency term in the phase response of the channel [29]. Hence, the PDD results in adding $-2\pi k n_\xi/N$ to the phase of CFR value of the k-th sub-carrier.

Phase Sanitization

In many prior work [20, 30], it is found that the AOA of the direct signal path has smaller variation compared with the indirect ones. Based on this, we classify the peak points obtained from the MUSIC algorithm into different clusters and the clusters with the smaller variance of AOA and TOA are selected as the candidates of direct signal path. Since the PDD varies for different packets, the variation of the TOA is difficult to be obtained. To solve this problem, the phase sanitization algorithm is proposed to avoid the impact of PDD on different packets.

By setting $\phi_{m,n}$ as the phase of CSI of the n-th sub-carrier at the m-th antenna, the aim of this approach is to optimize the value $\hat{\tau}_p$ to construct a linear fitting model for the phase of CSI to avoid the impact of PDD on different packets. The linear fitting function is defined as

$$\varphi(n) = 2\pi\Delta f(n-1)\mu + \beta \tag{13}$$

where n is the sub-carrier index. μ is the slope of the function. β is a constant. To optimize the linear fitting, we construct

$$\hat{\tau}_p = \arg\min_{\mu} \sum_{m}^{3} \sum_{n}^{30} (\phi_{m,n} - \varphi(n))^2 \tag{14}$$

To solve the optimization problem in Eq. (14), the well-known least square algorithm is employed to optimize the value $\hat{\tau}_p$ based on the CSI phase across 30 subcarriers at three antennas. Then, the phase of CSI can be modified into

$$\hat{\phi}_{m,n} = \phi_{m,n} + 2\pi\Delta f(n-1)\hat{\tau}_p \tag{15}$$

Gaussian Means Clustering

After the peak points are obtained from the MUSIC algorithm, the Gaussian means clustering algorithm is used to identify the direct signal path.

Specifically, first of all, k peak points are randomly selected as the k initial clustering centers, and then the K-means clustering is conducted to obtain clusters. Second, the distribution of the peak points in each cluster is examined as follows. If the peak points in a cluster do not approximately obey the Gaussian distribution, two random peak points in this cluster are selected as the two new clustering centers, and then the K-means clustering of peak points in this cluster is conducted to obtain two new clusters. This process is continued until the peak points in every cluster approximately obey the Gaussian distribution or cannot be split further.

3.3.3 Target Localization

To locate the target, the Second-order Cone Programming (SOCP) relaxation approach is used to transform the localization problem into a convex one based on the interior-point algorithm [31]. Specifically, by setting $\mathbf{ff}_1, \mathbf{ff}_2, \ldots, \mathbf{ff}_N$ and $\mathbf{p}$ ($\mathbf{p}, \mathbf{ff}_i \in R^2$, $i = 1, \ldots, N$) as the locations of the APs and target location, the distance between the target and each AP is estimated as

$$d_i = 10^{\frac{L_i}{10\gamma}}, \quad i = 1, \ldots, N \tag{17}$$

where $L_i = 10\log_{10}\frac{P_T}{P_i}$ (in dB). γ is the path loss exponent. P_T and P_i are the transmit power of the target and receive power at the i-th AP respectively.

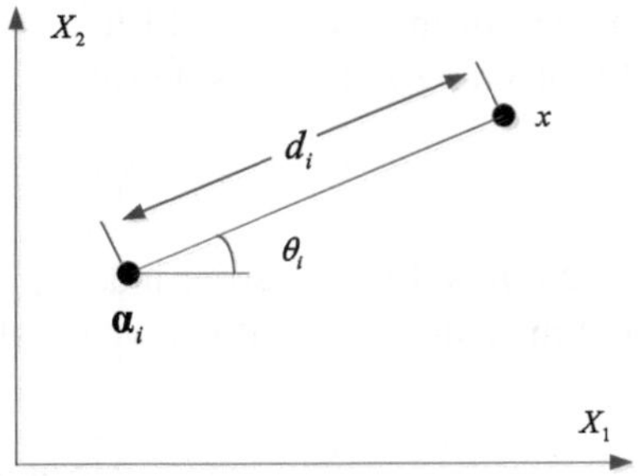

Fig. 9 Geometrical relations between the target and AP

Based on the geometrical relations in Fig. 9, the angle between the target and *i*-th AP is calculated by

$$\theta_i = \arctan\left(\frac{x_2 - \alpha_{i2}}{x_1 - \alpha_{i1}}\right) \tag{18}$$

where $\mathbf{x} = [x_1, x_2]$ and $\mathbf{ff}_i = [\alpha_{i1}, \alpha_{i2}]$ are the physical coordinates of the target and *i*-th AP. For simplicity, it is assumed that the APs and target are located with the same height. The impact of height difference between the APs and target on the AOA estimation will be discussed further in the following.

Based on the Least Square (LS) criterion, the estimated location of the target, $\hat{\mathbf{x}}$, can be obtained by minimizing the objective function below.

$$\hat{\mathbf{x}} = \arg\min_{\mathbf{x}} \sum_{i=1}^{N} (\|\mathbf{p} - \mathbf{ff}_i\| - d_i)^2 + \sum_{i=1}^{N} \left(\mathbf{c}_{\mathrm{i}}^{\mathrm{T}}(\mathbf{p} - \mathbf{ff}_i)\right)^2 \tag{19}$$

where $\mathbf{c}_{\mathrm{i}} = [-\tan(\theta_i), 1]$.

The problem in Eq. (19) is obviously non-convex due to the fact that the second derivative of Eq. (19) is greater than zero. To convert the problem in Eq. (19) into a convex one, the auxiliary variables are set as $r_i = \|\mathbf{x} - \mathbf{ff}_i\|$, $\mathbf{r} = [r_i]$, $\mathbf{z} = [z_i]$, and $\mathbf{g} = [g_i]$, where $z_i = \|\mathbf{x} - \mathbf{ff}_i\| - d_i$ and $g_i = \mathbf{c}_{\mathrm{i}}^{\mathrm{T}}(\mathbf{x} - \mathbf{ff}_i)$. Then, it is obtained that

$$\begin{aligned} s.t.\ r_i &= \|\mathbf{x} - \mathbf{ff}_i\| \quad i = 1, \ldots, N \\ r_i &= \|\mathbf{x} - \mathbf{ff}_i\| \quad i = 1, \ldots, N \\ z_i &= r_i - d_i, \quad i = 1, \ldots, N \\ g_i &= \mathbf{c}_{\mathrm{i}}^{\mathrm{T}}(\mathbf{x} - \mathbf{ff}_i), \quad i = 1, \ldots, N \end{aligned} \tag{20}$$

For the optimization, it is always assumed that the objective is a linear function of the variables. This can be done via the epigraph representation of the problem, which is based on adding a new scalar variable. Based on [32], the epigraph variables are set as t_1, t_2, and t_3, and then it is obtained that

$$
\begin{aligned}
&\underset{\mathbf{x},\mathbf{r},\mathbf{z},\mathbf{g},t_1,t_2}{\text{minimize}}\, t_1 + t_2 \\
&s.t.\ r_i \le \|\mathbf{x} - \mathbf{ff}_i\| \quad i = 1,\ldots,N \\
&\quad z_i = r_i - d_i, \quad i = 1,\ldots,N \\
&\quad g_i = \mathbf{c}_\mathrm{i}^\mathrm{T}(\mathbf{x} - \mathbf{ff}_i), \quad i = 1,\ldots,N \\
&\quad \left\| \begin{bmatrix} 2\mathbf{z} \\ t_1 - 1 \end{bmatrix} \right\| \le t_1 + 1, \quad \left\| \begin{bmatrix} 2\mathbf{g} \\ t_2 - 1 \end{bmatrix} \right\| \le t_2 + 1
\end{aligned} \tag{21}
$$

Based on Eq. (21), it can be found that the localization problem is transformed into a convex one, which can be solved by using the CVX package [33].

3.4 *Experimental Evaluation*

The three main results obtained from the experimental evaluation are summarized as follows.

First of all, the proposed system achieves the median localization error 0.7 m, which is lower than the one by the systems using the RSS or AOA measurement solely.

Second, the proposed AOA estimation approach achieves the median angle error within 5°, which is lower than the one by the conventional MUSIC algorithm.

Last but not least, the proposed direct signal path identification approach is with high identification precision, which enables to effectively reduce the large error probability.

3.4.1 Experimental Setup

Figure 10 shows the geometrical structure of the testbed. There are four APs and each of them is equipped with an Intel 5300 Wi-Fi NIC. The locations of APs are measured by using a laser range finder. Another AP which is configured in the transmission mode is selected as the target to transmit the signal. The Linux CSI toolkit [34] is applied to collect the CSI at each AP, and then transmit it to the location server. All the calculations are executed in the location server.

3.4.2 Performance of AOA and TOA Estimation

The test location P1 (with a black circle) in Fig. 10 is selected as an example to show the performance of the proposed AOA estimation approach in Fig. 11. For the comparison, the AOA at the AP1, AP2, AP3, and AP4 is 19°, 21°, −6°, and −19° respectively.

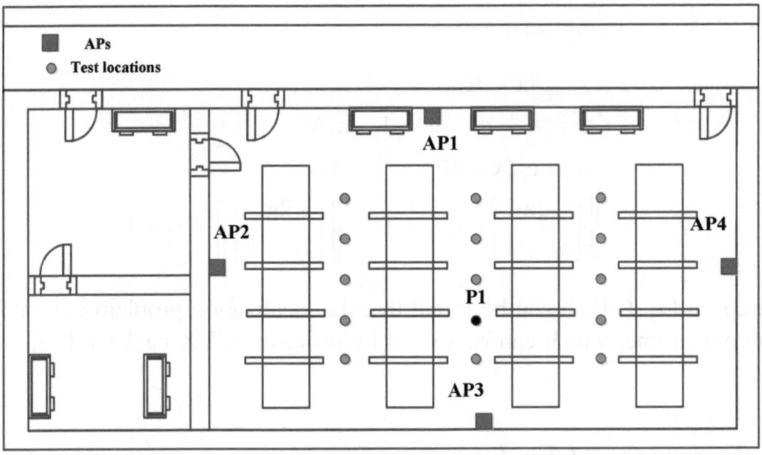

Fig. 10 Geometrical structure of the testbed

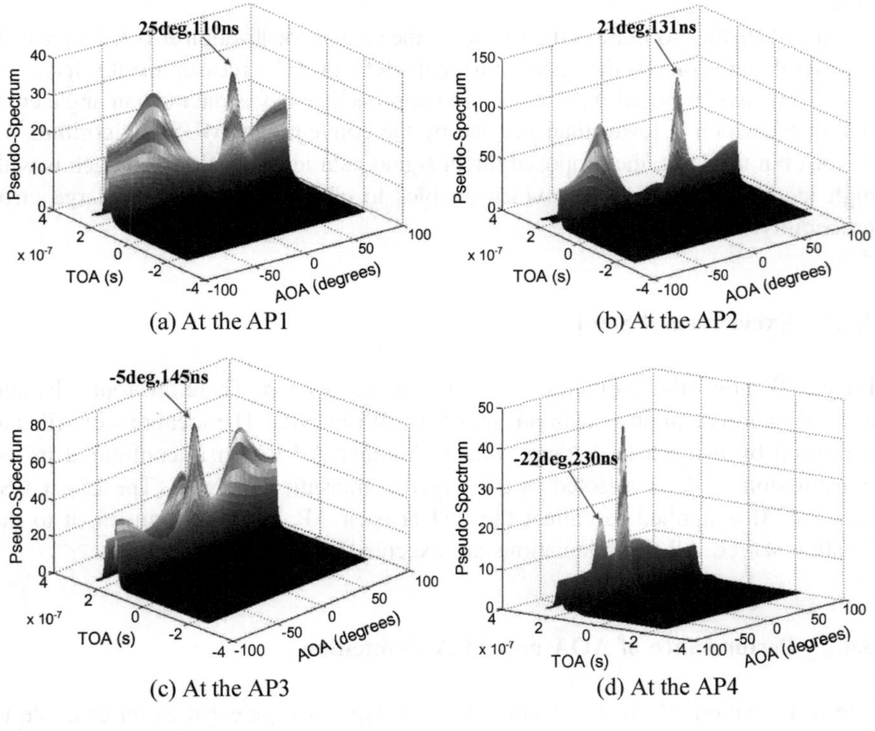

Fig. 11 Results of AOA and TOA estimation

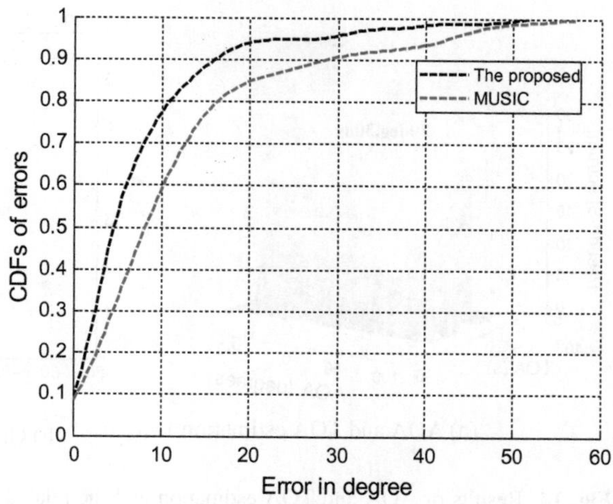

Fig. 12 CDFs of errors of AOA estimation

Figure 12 shows the CDFs of errors of AOA estimation by using the proposed and conventional MUSIC algorithms without spatial smoothing. The median and 80% errors by the proposed approach are about 4° and 11°, which are much smaller than the ones by the MUSIC algorithm.

3.4.3 Performance of Direct Signal Path Identification

Also by taking P1 as an example, Figs. 13, 14, 15 and 16 show the results of AOA and TOA estimation and the related Gaussian means clustering with phase sanitization at each AP. From these figures, we can find that the peak with the highest spectrum value is not always corresponding to the direct signal path since the direct signal path in the indoor environment is sometimes with the weaker RSS compared

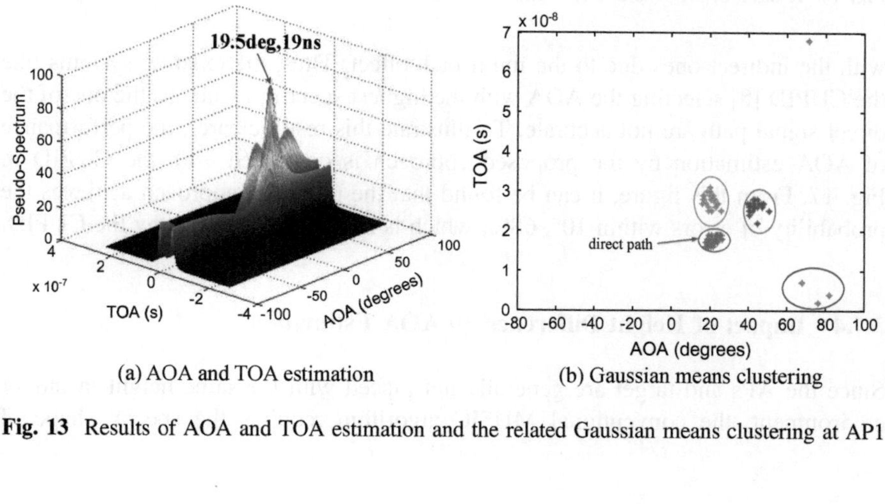

(a) AOA and TOA estimation (b) Gaussian means clustering

Fig. 13 Results of AOA and TOA estimation and the related Gaussian means clustering at AP1

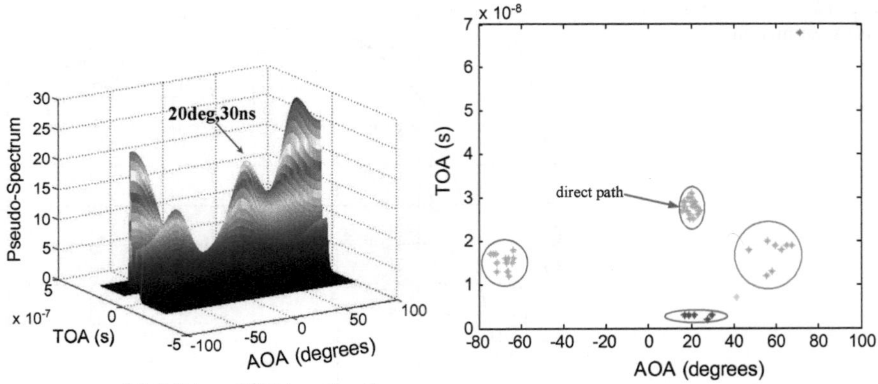

(a) AOA and TOA estimation (b) Gaussian means clustering

Fig. 14 Results of AOA and TOA estimation and the related Gaussian means clustering at AP2

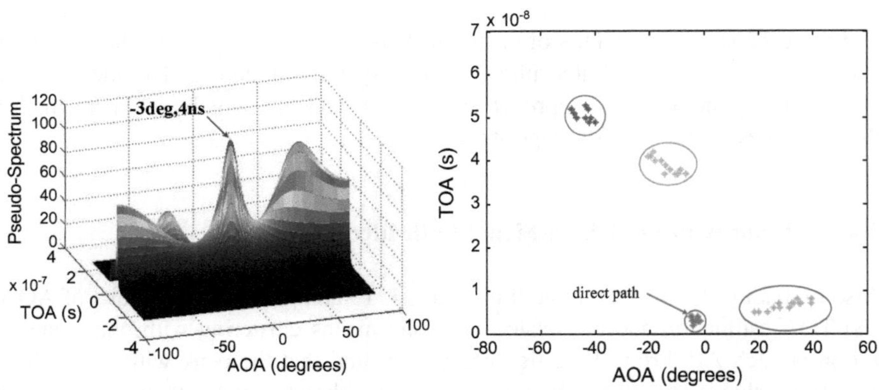

(a) AOA and TOA estimation (b) Gaussian means clustering

Fig. 15 Results of AOA and TOA estimation and the related Gaussian means clustering at AP3

with the indirect ones due to the multi-path effect. Thus, the existing systems like the CUPID [8] selecting the AOA with the highest spectrum value as the one of the direct signal path are not accurate. To illustrate this result clearer, the performance of AOA estimation by the proposed approach is compared with the CUPID in Fig. 17. From this figure, it can be found that the proposed approach achieves the probability of errors within 10°, 68%, which is higher than the one by the CUPID.

3.4.4 Impact of Height Difference on AOA Estimation

Since the APs and target are generally not placed with the same height in indoor environment, the conventional MUSIC algorithm requires the special shape of

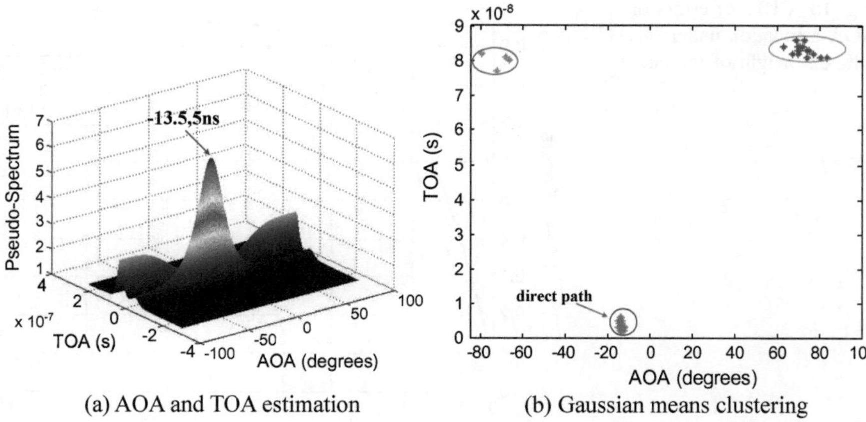

Fig. 16 Results of AOA and TOA estimation and the related Gaussian means clustering at AP4

Fig. 17 CDFs of errors of AOA estimation

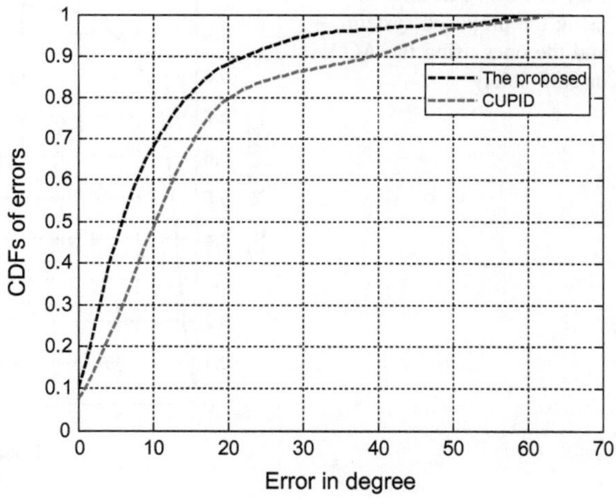

antenna array like the L-shape [35] for the AOA estimation. However, this system is able to overcome the problem of the requirement of special shape of antenna array by using multiple APs for the localization. Figure 18 shows the CDFs of errors of AOA estimation by placing the APs with 2.7 m height and target with 2.7, 2.1, and 1.5 m height respectively. From this figure, it is observed that when the APs and target are placed with the same height, the median and 70% errors are within 5° and 10°. With the increase of height difference between the APs and target, the error of AOA estimation increases as expected.

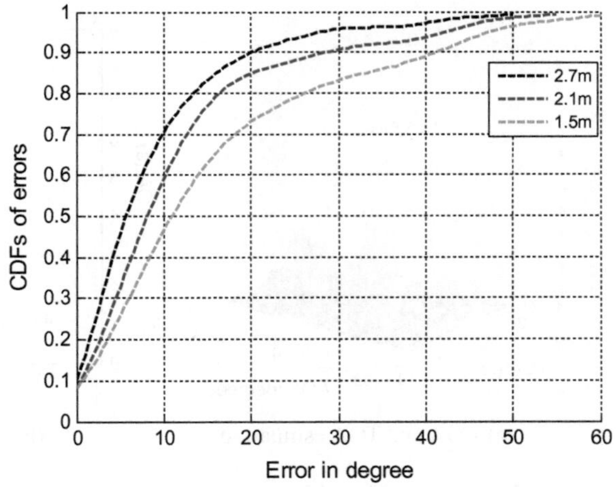

Fig. 18 CDFs of errors of AOA estimation under different height of the target

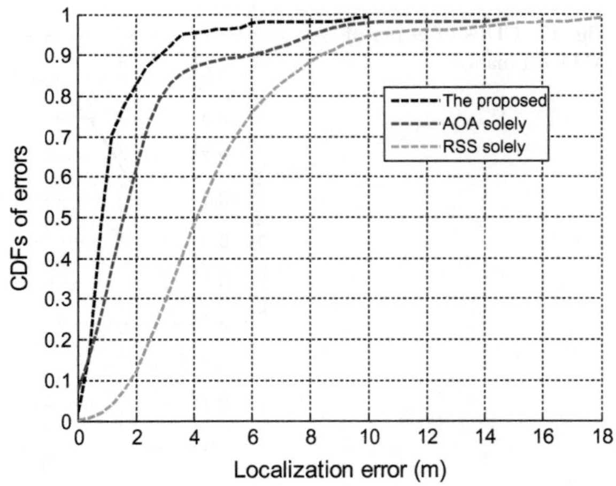

Fig. 19 CDFs of errors by using the proposed system and the ones using the AOA or RSS solely

3.4.5 Location Estimation

In this system, the target sends 200 signal packets in each second to the APs operating in the monitor mode, and then the APs transmit the CSI to the location server. After that, the previous 30 packets are used to form a dataset for the testing. To examine the localization performance of this system, it is compared with the existing ones by using the AOA [2] or RSS [13] measurement solely.

As shown in Fig. 19, the median error by the proposed system is 0.7 m, which is smaller than the one by the systems using the AOA or RSS measurement solely. Thus, this system is verified to be able to achieve the sub-meter localization accuracy without any special hardware modification.

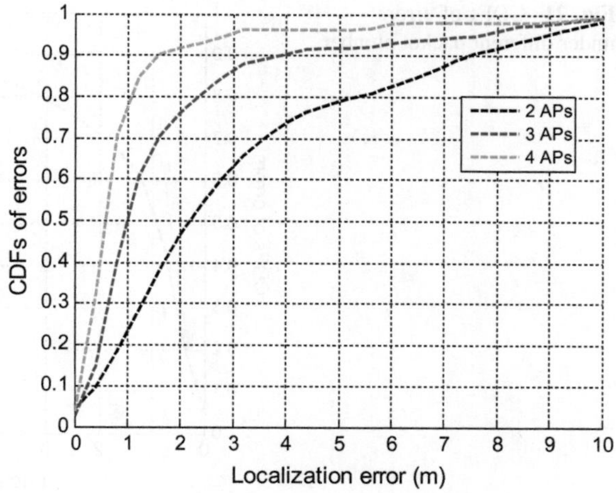

Fig. 20 CDFs of errors under different AP number

3.4.6 Impact of AP Number

To investigate the impact of AP number on the localization accuracy of the proposed system, Fig. 20 shows the CDFs of errors as the AP number increases from 2 to 4. As expected, the increase of AP number improves the localization accuracy. For example, the median error is 2, 1, and 0.7 m under the two, three, and four APs conditions respectively.

3.4.7 Impact of Packet Number

Since this system uses multiple signal packets to identify the Line-of-sight (LOS) and Non-line-of-sight (NLOS) signal paths, the packet number probably affects the precision of direct signal path identification. On account of this, Fig. 21 shows the CDFs of errors as the packet number increases from 10 to 50. From this figure, it can be found that the localization accuracy changes slightly with the variation of packet number. For example, as the packet number increases from 10 to 50, the median error decreases from 0.8 to 0.64 m.

3.4.8 System Latency

To evaluate the system latency on the localization, the time cost is investigated with different number of packets and APs. From Fig. 22, it is observed that as the packet number used for the localization increases from 10 to 40, the time cost increases from 12.5 to 45.2 s when using 2 APs for the localization. The more number of APs used for the localization, the higher time cost needed for the data processing.

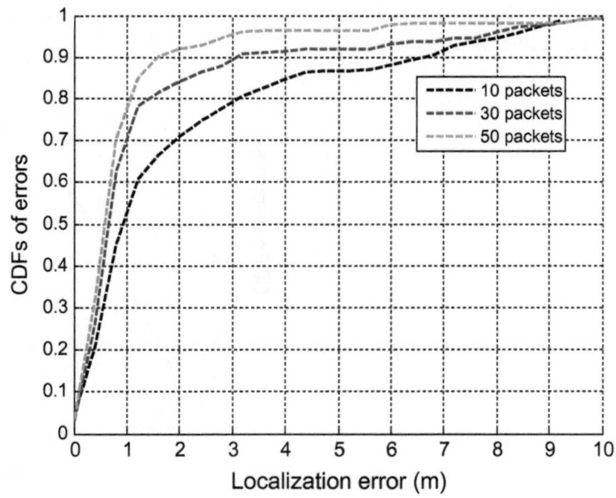

Fig. 21 CDFs of errors under different packet number

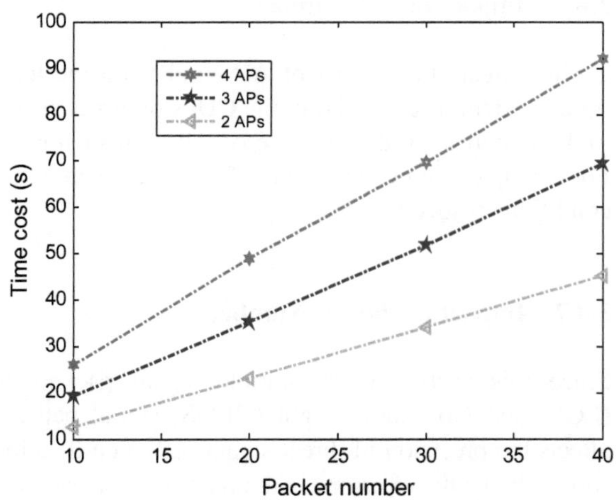

Fig. 22 Time cost with different number of packets and APs

As expected, the increase of AP number costs more time to locate the target. For example, when using 20 packets, the time cost is 23.1, 35.2, and 48.7 s under the two, three, and four APs conditions.

4 LOS/NLOS Identification for OFDM Based Localization

4.1 Introduction

Recently, the MIMO with smart antennas has been applied widely in current wireless communication system such as IEEE 802.11n, LTE, WiMAX, and et al.

The MIMO is so prevalent mainly due to its capacity of improving the data throughput and spectral efficiency. At the same time, the use of MIMO antennas at both two link ends can combat Radio Frequency (RF) interference by designing the beamforming, i.e., focusing the power for desired direction. An important feature of MIMO channel is the occurrence of multipath components. So, the MIMO provides us great potentials to develop vast number of MIMO transmission techniques and intelligent applications, such as direction finding [4], fine-gained active and passive indoor localization [7, 8, 36], and et al.

Besides the active and passive localization, there are vast computing applications, including adaptive communication [37], human activity recognition [38], location authentication [39], and et al. The presence of NLOS propagation, which happens when the direct path between transmitter and receiver ends is completely blocked, degrades the performance of OFDM based localization systems working in the complex indoor environment.

Therefore, knowing the availability of LOS propagation is significant important. In the wireless security field, awareness of the propagation environment of Mobile Terminal (MT) can obtain an additional physical defender. As shown in Fig. 23, the network administrator of a coffee shop can prevent the attacker from eavesdropping on the traffic of authorized users or injecting adverse traffic depending on whether the mobile terminal lies in the LOS. Besides network security field, the effective power management and reliable communication of Base Station (BS) and Access Point (AP) can be realized easily by sensing ambient environment.

For the fine-gained localization systems using Time-of-flight (TOF) or AOA, handling the NLOS environment is challenging. Although the direct path (or called the first arriving path) can be retrieved correctly, the distance corresponding to the estimated TOF is normally longer than the actual distance between the two ends, and the AOA measurement is normally spurious compared to the real one. Thus, the NLOS propagation degrades localization accuracy. Therefore, the ability of sensing propagation condition is the domain factor affecting the positioning accuracy. Even for the coarse-gained fingerprint-based localization system, the positioning accuracy is related to propagation state since the RSS suffers from severe fluctuation and multipath effect under the NLOS condition. The pioneer Radar [13] constructs the

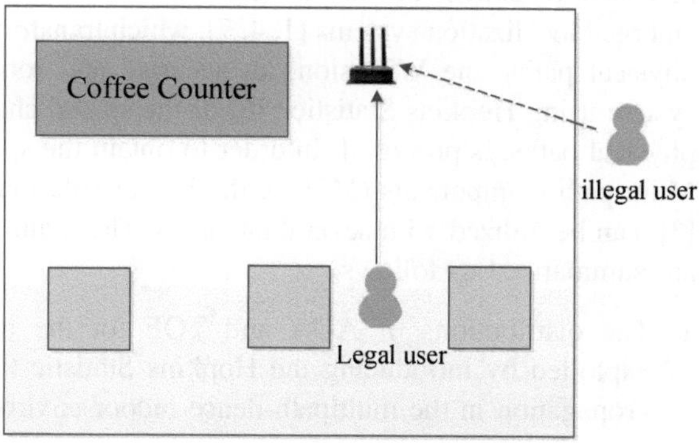

Fig. 23 An example of authentication by the availability of LOS propagation

radio maps under the LOS and NLOS conditions respectively, which can combat the accuracy degradation by recognizing propagation condition.

In summary, compared to the LOS environment, the NLOS propagation, which involves enormous reflectors and scatters, induces RSS fluctuation and an extra TOF and AOA bias. Consequently, a large number of methods emerge to mitigate the NLOS impact on the system performance. Tracking moving target [40] utilizes a filtering algorithm to decrease NLOS measurement errors and achieve high accuracy, while the error statistics [41] are used for NLOS error reduction. Differing from the method striving better parameters from the NLOS-distort measurement, an alternative way to cope with NLOS propagation is to recognize NLOS environment first, and then discard the corrupted measurement [42, 43]. In this circumstance, the LOS identification is the key component, which will be further discussed in the following.

The LOS identification has been studied extensively in recent years. The representative LOS identification system [43] uses the precise CIR to calculate the kurtosis, mean excess delay spread, and root mean square delay spread. But this system cannot be applied in current Wi-Fi or LTE network due to the bandwidth limitation. Recently, LiFi [44] realizes the LOS identification on the commodity Wi-Fi devices by using the CSI, but it just uses the CSI amplitude. Later Phase U [45] improves the performance by exploiting both amplitude and phase of the CSI. Those systems indeed solve the LOS identification problem to a certain extend, but in practical it is found that they are sensitive to the system parameters setting, such as channel selection, which may vary for different users in Wi-Fi or LTE network. For instance, the Wi-Fi at 2.4 and 5.2 GHz has many channels with independent center frequencies. Thus, it is difficult to select the unified threshold for various system parameters configuration.

The wireless signals working on different channels experience frequency-dependent attenuation and cause frequency dependent phase under the same TOF. Consequently, the fixed threshold used in [44, 45] is not decent for different channel configuration. The intuition is that the channel measurement, such as the CIR and CSI, differs with frequency, but the physical paths remain unchanged. So, the characteristics of frequency-invariant physical paths can be exploited to distinguish the LOS and NLOS environment. However, the problem is that how to extract the useful features of physical path? Inspired by the recent emerged localization systems [1, 4, 7], which transform the channel measurement to physical paths, the Wi-vision, an accurate and robust LOS/NLOS identification system using Hopkins Statistic [46] of the spatial characteristics for the underlying physical paths, is proposed. In order to obtain the spatial information of individual Multi-path Components (MPCs), the higher solution approaches, such as MUSIC [3], can be utilized with several antennas. The main contributions of the Wi-vision are summarized as follows.

1. The distributions of AOA and TOF for the underlying physical paths are exploited by introducing the Hopkins Statistic to distinguish LOS and NLOS propagation in the multipath-dense indoor environment.

2. The Wi-vision using Intel 5300 commodity Wi-Fi NIC is implemented in various indoor sceneries and demonstrated to be featured with high environment sensing ability. The experimental results show that the Wi-vision is able to achieve an overall LOS and NLOS detection rates, 91 and 89%, and its accuracy is stable with the change of system configuration.
3. Wi-vision is expected to achieve better performance on the current LTE and future 5G cellular networks since enormous MIMO antennas will be equipped on the BS.

4.2 *Preliminaries*

4.2.1 Challenge of System Robustness

In Wi-Fi network, the AP can be configured to work on many different channels and the operated channel has different center frequency. The robustness is a challenge for the applications working on different parameters configuration [47].

A. *Motivation of system design*

In fact, the wireless channel faithfully records how the signals traverse along the physical paths between the two link ends. By assuming that there is a single path signal with narrow band, the channel can be describe as

$$h = \alpha e^{j2\pi f\tau} \tag{22}$$

where f is the signal frequency and τ is the TOF taken by the signal from the transmitter to receiver, α is the attenuation factor. By considering the practical case in which the signals traverse along N paths, the channel measurement at the receiver which is equipped with a linear antenna array consisting of M elements can be written as

$$h_m = \sum_{i=1}^{N} \alpha_i e^{-j2\pi f\tau_i} e^{-j2\pi(m-1)\frac{d}{\lambda}\tau_i \sin\theta_i} \tag{23}$$

where h_m is the channel measurement at the m-th antenna and α_i is the attenuation of the i-th path, c is the light speed, τ_i and θ_i are the TOF and AOA of the i-th path respectively. From Eq. (23), it is observed that the AOA and TOF depend on the frequency that is to say although the channel measurement varies across different frequency, the physical paths remain unchanged. Thus, the channel can be described by the frequency-independent attributes of signals, i.e., AOA and TOF. Consequently, the two attributes are naturally configuration-independent. Utilizing AOA and TOF demonstrates great advantages over the schemes which employ the raw phase and amplitude. Thus, the problem is that how to resolve the MPCs at receiver and extract effective features to differentiate the LOS and NLOS environment?

4.2.2 System Design

System Overview

In this section, the way to distinguish the LOS and NLOS scenarios by using the Wi-Fi network will be described in detail. The framework of Wi-vision consists of two components, including the (1) feature extraction and data training and (2) LOS and NLOS environment identification.

(1) **Feature extraction and data training**. In this component, first of all, Wi-vision resolves the MPCs and then calculates the Hopkins statistic under LOS and NLOS condition respectively. Second, Wi-vision relies on the optimal threshold to differentiate LOS and NLOS by using the Probability Density Function (PDF) with respect to Hopkins statistic.
(2) **LOS and NLOS identification**. After resolving the MPCs, the peak points in a 2-dimension (2D) space are detected to calculate the corresponding Hopkins statistic for different environment, and then the environmental type is determined based on the optimal threshold.

Phase Calibration and MPCs Resolving

Many prior indoor localization systems [1, 7] have realized AOA and TOF estimation on Wi-Fi devices. Due to the hardware and bandwidth limitation, most of them cannot be deployed widely in current Wi-Fi network. The SpotFi [7] proposes to expand the number of antennas by reconstructing the CSI measurement matrix. Thus, up to now, the number of antennas is not the dominant hamper, especially for the LTE and massive MIMO system since a larger number of antennas are available.

Before jointly estimating the AOA and TOF by leveraging the smoothing MUSIC algorithm in [7], the CSI phase should be sanitized. In [48, 49], various phase preprocessing approaches have been discussed, the common purpose is to eliminate the phase error caused by Sample Time Offset (STO) and PDD. In fact, in IEEE 802.11n standard, the 40 MHz channel consists of two 20 MHz sub-channel, so the unwrapped phase of the two marginal subcarriers exist big jump, as illustrated in [7]. Consequently, the estimated slope across 30 subcarriers is relative large, which may result in overestimation. Thus, the CSI phase of subcarriers in each sub-channel needs to be unwrapped. From [49], the phase model of each subcarrier in each packet can be describe as

$$\hat{\varphi}(n) = \varphi(n) + 2\pi n \Delta f \mu + \beta \tag{24}$$

where $\varphi(n)$ is the true CSI phase of the n-th subcarrier and β is the phase noise. The least square approach is used to estimate the slope as follows.

$$\hat{\mu} = \arg\min_{\mu} \sum_{m,i,n=1}^{M,K,N} (\phi_i(m,n) - \hat{\varphi}(n))^2 \tag{25}$$

where $\phi_i(m,n)$ is the unwrapped phase of the CFR of the n-th subcarrier in the i-th sub-channel received at the m-th antenna. Here, the operation $\phi_i(m,n) - n\hat{\mu}$ removes the phase error in different packets. In essence, the slop $\hat{\mu}$ is a large additional time delay for all the paths. Although this time delay can be estimated, removing the phase error completely is impossible that is to say the estimated TOF is not the real one but the relative TOF is reserved. Fortunately, this system utilizes the variance of TOF instead of the actual TOF, so the relative TOF is still beneficial.

Feature Extraction

The studies in [4, 7] have found that the NLOS paths behave more randomly compared to the LOS ones, but they cannot capture the concrete behavior since the variation of phase and amplitude is coarse-granularity representation. Prior related studies in [44, 45] mainly exploit the statistical features that the randomness of LOS and NLOS propagation demonstrates significant difference. In fact, the accuracy of parameters estimation is affected by the Signal-to-Noise Ratio (SNR), so the spatial parameters are stable and accurate for the parameters estimation when the MT is in LOS. Contrary to the LOS scenario, the estimated parameters have fluctuation in the NLOS scenario with low SNR. In the following, the performance of MUSIC algorithm will be demonstrated as a function of SNR under the conditions that there are 8 signals with the AOA −30°, −25°, −20°, 30°, 20°, 10°, 0°and TOF 5, 15, 25, 35, 45, 55, 65 and 80 ns respectively. The parameters estimation is conducted for 20 times, and then all the peak points in the 2D space are plotted in Fig. 24. From this figure, it can be observed that the estimated parameters deviate from the real ones as the SNR varies from 20 to −10 dB. Therefore, when the MT is in the NLOS, the estimated parameters stay loosely in the 2D space. At the same time, when there are persons walking around, the spatial parameters can be added as an additional variation.

Figure 25 shows the testbed, in which there are normal activities such as walking around in this environment. The peak points in the 2D space are plotted in Fig. 26. In the LOS environment, there are several natural clusters and each of them indicates a group of MPCs with similar AOA and TOF. Contrary to the LOS, there is no distinct natural cluster under the NLOS environment and all the peak points exhibit large randomness. The key features are mainly derived from two aspects. On the one hand, the LOS path is naturally stable since there is no obstacle between

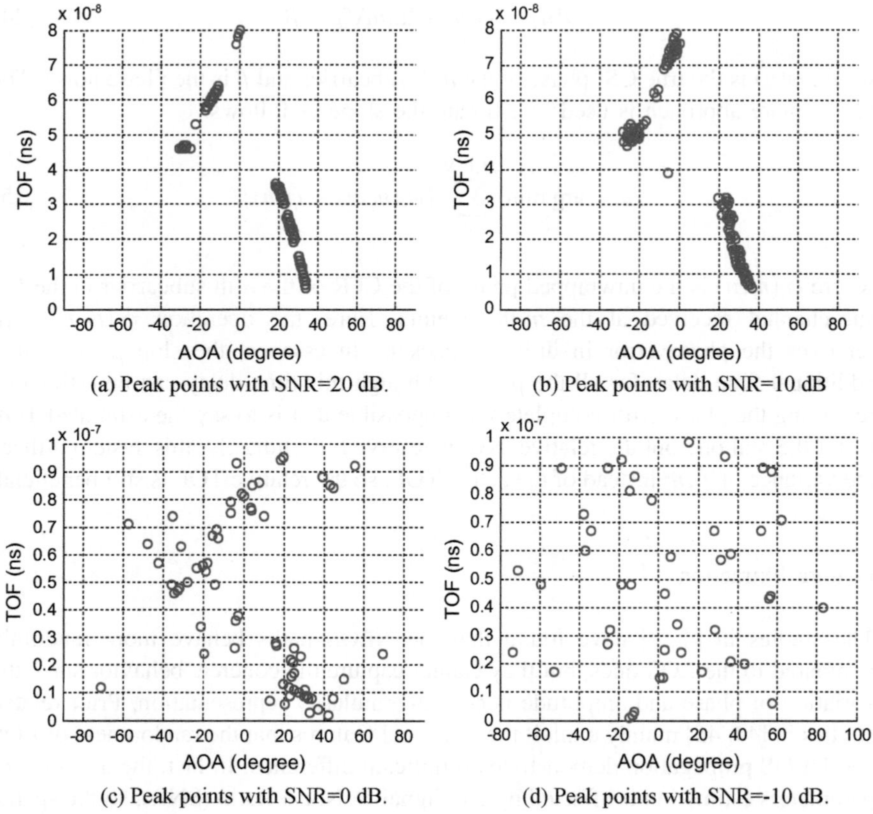

(a) Peak points with SNR=20 dB. (b) Peak points with SNR=10 dB

(c) Peak points with SNR=0 dB. (d) Peak points with SNR=-10 dB.

Fig. 24 Distribution of peak points with different SNR

the AP and client and even the dominant reflectors demonstrate to be relatively stable compared to the NLOS paths. On the other hand, the NLOS signal is suffered from fierce attenuation and the underlying physical paths traversed by signals are random. Thus, it is found that the path distribution is different under the LOS and NLOS conditions. Based on this, the measure of clustering tendency is exploited to construct the LOS and NLOS identification scheme.

(1) **Measure of Clustering Tendency**. The aim of studying clustering tendency is to identify whether a given data set has predisposition to be partitioned into natural clusters. Despite a data set without inherent natural clusters can be clustered, the clustering result is worthless. Specially, the 2D data set collected under NLOS condition cannot be clustered since the LOS path does not exist. Thus, the Hopkins statistic [46] is adopted to measure the clustering tendency.

After collecting the peak points estimated from several packets in a certain environment, a normalized data set is constructed as $S = \{s_i | i = 1, \ldots, M\}$ in the 2D space and $s_i = (\theta_i, \tau_i)$. By letting $P = \{p_j | j = 1, \ldots, N,\ N < M\}$ be the collection of sampling points randomly placed in the 2D space, sampling collection $Q = \{q_i | i = 1, \ldots, N, N < M\}$ be a subspace of the whole 2D data space, and P be the collection of sampling points randomly placed in S, the minimum distance from $p_i (\in P)$ to the points in S is calculated by

$$x_i = \min_{v \in S} \{dist(p_i, v)\} \tag{26}$$

In addition, the minimum distance from $q_i (\in Q)$ to the points in S is calculated by

$$y_i = \min_{v \in S,\, v \neq q_i} \{dist(q_i, v)\} \tag{27}$$

Thus, the Hopkins statistic equals to

$$h = \frac{\sum_{i=1}^{N} y_i}{\sum_{i=1}^{N} x_i + \sum_{i=1}^{N} y_i} \tag{28}$$

Hopkins statistic h distributes in different region when the data set demonstrates different distributions. Here, the value h approaches 0.5 and 0 when the distribution of date set S is uniform and naturally clustered respectively in the 2D space. Nonetheless, this criterion cannot be used directly since the underlying signal paths do not obey these two distributions accurately under the NLOS and LOS conditions. Therefore, the optimal threshold needs to be set to differentiate these two conditions.

(2) **Optimal Identification Threshold**. Given a collection of peak points, the LOS/NLOS identification problem can be described as a binary hypothesis test problem with the LOS case H_0 and NLOS case H_1, as shown below.

$$\begin{cases} H_0 : h > H_{th} \\ H_1 : h < H_{th} \end{cases} \tag{29}$$

where H_{th} is a threshold for Hopkins statistic and the optimal threshold H_{th} can be calculated by employing methods used in [50].

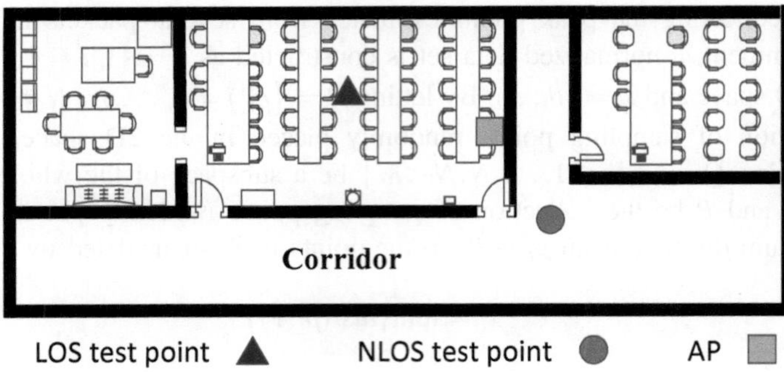

Fig. 25 Layout of testbed

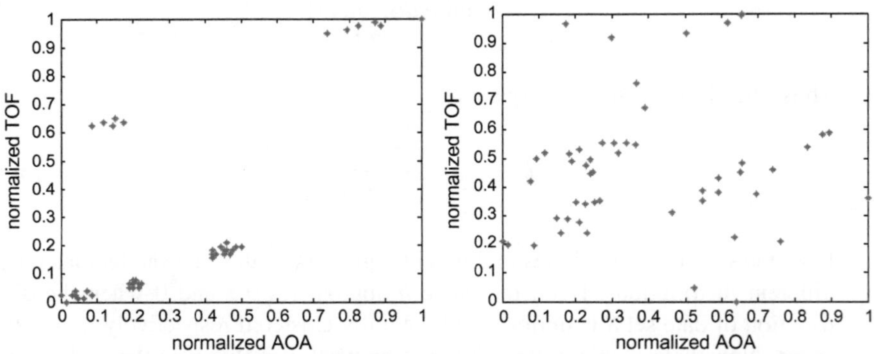

(a) Peak points under the LOS environment (b) Peak points under the NLOS environment

Fig. 26 Distribution of peak points under the LOS and NLOS environment

4.3 Experimental Results

A. *Experiment Methodology*

(1) **Experimental Layout**. The ProBox23 MS-B083 mini PC equipped with the Intel 5300 Wi-Fi NIC with three antennas and Samsung S4 smart phone are selected as the AP and client respectively. The PC runs the Ubuntu 14.0 Operating System and uses the Linux CSI tool [51] to extract the CSI from each packet. As shown in Fig. 27, the data collection is conducted on a floor which consists of one corridor and several laboratories and office rooms (Fig. 27).
(2) **Evaluation Metrics**. Two metrics are used to evaluate the Wi-vision, including the (1) LOS detection rate, i.e., the percentage of identifying the

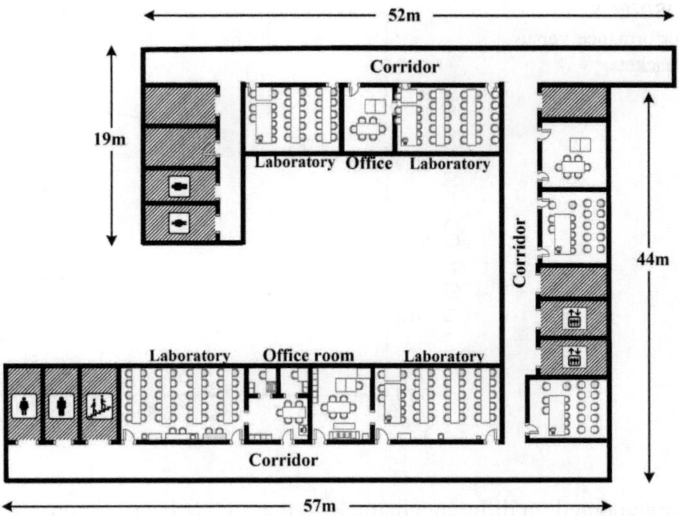

Fig. 27 Experimental layout

LOS under the LOS condition. (2) NLOS detection rate, i.e., the percentage of identifying the NLOS under the NLOS condition.

B. *Performance Evaluation*

(1) **Overall Performance on Different Channels**. The AP is configured to operate on different channels and then the CSI data are collected under the LOS and LOS conditions. In order to obtain a large difference of configuration, the channel in the 2.4 and 5 GHz frequency bands are selected. Due to the phase ambiguity in the 2.4 GHz band [4], the Wi-vision is

Fig. 28 LOS/NLOS detection performance on different channels

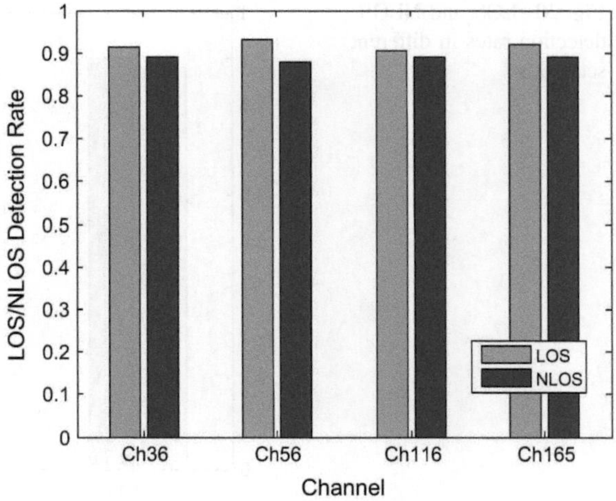

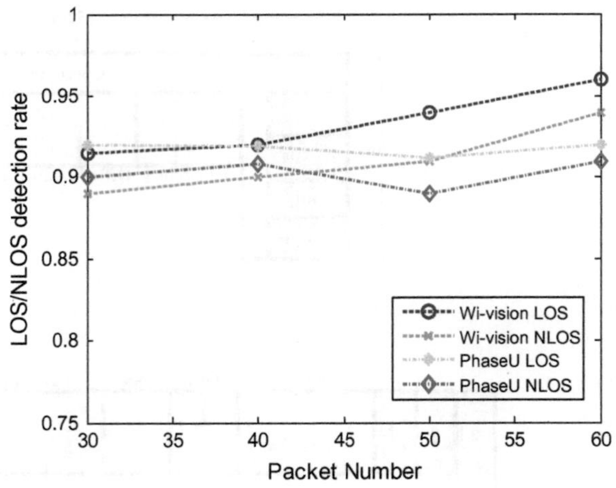

Fig. 29 LOS/NLOS detection performance versus number of packets

implemented on different channels in the 5 GHz band. Figure 28 shows the LOS/NLOS identification performance by using 30 packets. From this figure, it is found that the Wi-vision achieves LOS and NLOS detection rates above 90 and 87% respectively. In all, the proposed system utilizing the frequency-independent features of underlying physical paths is able to achieve good and consistent performance under both the LOS and NLOS conditions.

(2) **Impact of Number of Packets**. More packets normally result in more distinct spatial features of underlying physical paths. Figure 29 shows the variation of detection rates, the figure show that the LOS and NLOS detection rates are increased with the increase of the number of packets.

(3) **Impact of Environmental Structure**. The performance of AOA estimation significantly depends on the environmental structure. To illustrate this

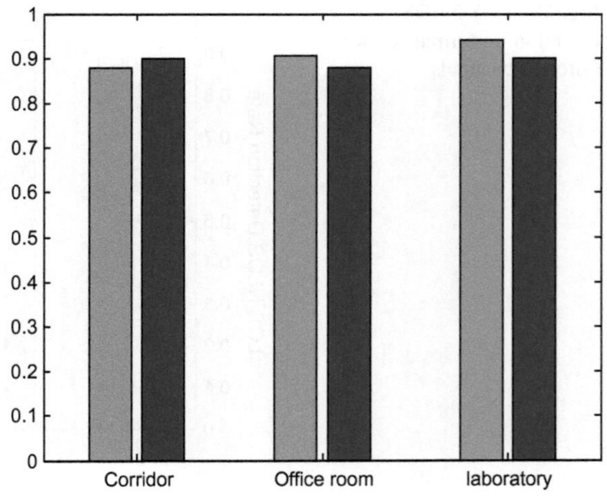

Fig. 30 LOS and NLOS detection rates in different scenarios

result clearer, the data collection is conducted in different environments, including the corridor, office rooms, and laboratories. Figure 30 shows the LOS and NLOS detection rates in various environments, from which it is found that there is a slight performance variation for the Wi-vision. The lowest LOS detection rate 88.2% is obtained in the corridor, which can be interpreted by the fact that resolving the MPCs is difficult in the narrow environment where many correlated signals exist.

5 Conclusions

The proposed system uses the commodity Wi-Fi devices with three antennas to precisely estimate the AOA between the APs and target, and eventually achieve the sub-meter localization accuracy without any special hardware modification. By integrating the AOA and RSS measurements to locate the target, the localization performance of this system is significantly improved compared with the conventional ones by using the AOA or RSS measurement solely. Moreover, an accurate and robustness LOS identification system is proposed. First of all, the CSI phase is calibrated by using a sanitization algorithm to resolve the MPCs. The Hopkins statistic is introduced to measure the features of distribution of AOA and TOF under the LOS and NLOS condition. After that, the PDFs with respect to the Hopkins statistic in LOS and NLOS cases are used to construct an objective function, and then we calculate the optimal threshold. The experiment results shows that Wi-vision can achieve consistent LOS and NLOS detection rate of above 90% and 89% respectively while configuring different system parameters. Due to the hardware limitation, we implement Wi-vision on the commodity WiFi device. Our system can be easily implemented on the LTE and future 5G networks with advantages of MIMO antennas.

References

1. Kumar S, Gil S, Rus D (2014) Accurate indoor localization with zero start-up cost. In: Proceedings of the 20th annual international conference on mobile computing and networking, pp 483–494
2. Kawauchi K, Miyaki T, Rekimoto J (2009) Directional beaconing: a robust WiFi positioning method using angle-of-emission information. In: Proceedings of the 4th international symposium on location and context awareness, pp 103–119
3. Schmidt RO (1986) Multiple emitter location and signal parameter estimation. IEEE Trans Antennas Propag 34(1):276–280
4. Gjengset J, Xiong J, Mcphillips G, Jamieson K (2014) Phaser: enabling phased array signal processing on commodity WiFi access points. In: Proceedings of the 20th annual international conference on mobile computing and networking, pp 153–163

5. Tian Z, Tang X, Zhou M, Tan Z (2013) Fingerprint indoor positioning algorithm based on affinity propagation clustering. Eur J Wirel Commun Netw 2013(1):1–8
6. Chintalapudi K, PadmanabhaIyer A, Padmanabhan VN (2010) Indoor localization without the pain. In: Proceedings of the 16th annual international conference on mobile computing and networking, pp 173–184
7. Kotaru M, Joshi K, Bharadia D, Katti S (2015) SpotFi: decimeter level localization using WiFi. In: Proceedings of the 2015 ACM conference on special interest group on data communication. SIGCOMM 45(5):269–282
8. Sen S, Lee J, Kim KH, Congdon P (2013) Avoiding multipath to revive in building WiFi localization. In: Proceedings of the 11th annual international conference on mobile systems, applications, and services, pp 249–262
9. Yu K (2007) 3-D localization error analysis in wireless networks. IEEE Trans Wirel Commun 6(10):3473–3481
10. Wang S, Jackson BR, Inkol R (2014) Hybrid RSS/AOA emitter location estimation based on least squares and maximum likelihood criteria. In: Proceedings of the 26th Biennial symposium on communications, pp 24–29
11. Paolo B, Stefano L, Stefano C, Gaetano G (2009) A novel approach to indoor RSSI localization by automatic calibration of the wireless propagation model. In: Proceedings of the 2009 vehicular technology conference, pp 1–5
12. Youssef M, Agrawala A (2005) The horus WLAN location determination system. In: Proceedings of the 3rd international conference on mobile systems, pp 205–218
13. Bahl P, Padmanabhan, VN (2000) Radar: an in-building RF-based user location and tracking system. IEEE INFOCOM, 775–784
14. Li X (2006) RSS-based location estimation with unknown path loss model. IEEE Trans Wirel Commun 5(12):3626–3633
15. Mazuelas S, Bahillo A, Lorenzo RM, Fernandez P, Lago FA, Garcia E, Blas J, Evaristo JA (2009) Robust indoor positioning provided by real-time RSSI values in unmodified WLAN networks. IEEE J Sel Top Signal Process 3:821–831
16. Pajovic M, Orlik P, Koike-Akino T, Kim KJ, Aikawa H, Hori T (2015) An unsupervised indoor localization method based on received signal strength RSS measurements. In: Proceedings of the 2015 IEEE global communications conference (GLOBECOM). San Diego, CA, USA, 6–10 Dec 2015, pp 1–6
17. Tian Z, Fang X, Zhou M, Li L (2015) Smartphone-based indoor integrated WiFi/MEMS positioning algorithm in a multi-floor environment. Micromachines 6:347–363
18. Tian Z, Zhang Y, Zhou M, Liu Y (2014) Pedestrian dead reckoning for MARG navigation using a smartphone. Eur J Adv Signal Process 1:1–9
19. Mariakakis AT, Sen S, Lee J, Kim KH (2014) SAIL: single access point-based indoor localization. In: Proceedings of the 12th annual international conference on mobile systems, applications, and services. Bretton Woods, NH, USA, 16–19 June 2014, pp 315–328
20. Xiong J, Jamieson K, ArrayTrack: a fine-grained indoor location system. In: Proceedings of the 10th Usenix symposium on networked systems design and implementation, Lombard, IL, USA, 2–5 Apr 2013, pp 71–84
21. Qian K, Wu C, Yang Z, Zhou Z, Wang X, Liu Y (2016) Tuning by turning: enabling phased array signal processing for WiFi with inertial sensors. In: Proceedings of the 35th annual IEEE international conference on computer communications. San Francisco, CA, USA, 10–14 Apr 2016, pp 1–9
22. Zheng ZW (2010) Channel estimation and channel equalization for the OFDM-based WLAN systems. In: Proceedings of the 2010 international conference on e-business and e-government (ICEE), Guangzhou, China, 7–9 May 2010, pp 1691–1694
23. Bello P (1963) Characterization of randomly time-variant linear channels. IEEE Trans Commun Syst 11:360–393

24. Zhou Z, Yang Z, Wu C, Sun W (2014) LiFi: line-of-sight identification with WiFi. In: Proceedings of the 2014 proceedings IEEE INFOCOM, Toronto, ON, Canada, 27 Apr–2 May 2014, pp 2688–2696
25. Laxmikanth P, Surendra L, Ratnam DV, Babu SS (2015) Enhancing the performance of AOA estimation in wireless communication using the MUSIC algorithm. In: Proceedings of the 2015 international conference on signal processing and communication engineering systems, Vijayawada, India, 2–3 Jan 2015, pp 448–452
26. Shan TJ, Wax M, Kailath T (1985) On spatial smoothing for direction-of-arrival estimation of coherent signals. IEEE Trans Acoust Speech Signal Process 33:806–811
27. Shen J, Molisch AF (2012) Indirect path detection based on wireless propagation measurements. IEEE Trans Wirel Commun 11:4482–4493
28. Venkatraman S, Caffery J, You HR (2004) A novel ToA location algorithm using LoS range estimation for NLoS environments. IEEE Trans Veh Technol 53:1515–1524
29. Sen S, Radunovic B, Choudhury RR, Minka T (2012) You are facing the Mona Lisa: spot localization using PHY layer information. In: Proceedings of the 10th international conference on mobile systems, applications, and services, Lake District, UK, 25–29 June 2012, pp 183–196
30. Joshi K, Hong S, Katti S (2013) PinPoint: localizing interfering radios. In: Proceedings of the 10th Usenix conference on networked systems design and implementation. Lombard, IL, USA, 2–5 Apr 2013, pp 699–701
31. Boyd S, Vandenberghe L (2004) Convex optimization. Cambridge University Press, New York, NY, USA
32. Boyd SP (1998) Applications of second-order cone programming. Linear Algebra Appl 284:193–228
33. Grant M, Boyd S (2016) CVX: matlab software for disciplined convex programming. Available online: http://cvxr.com/cvx. Accessed on 15 June 2016
34. Halperin D, Hu W, Sheth A, Wetherall D (2011) Tool release: gathering 802.11n traces with channel state information. Comput Commun Rev 41:53
35. Tayem N, Kwon HM (2005) L-shape 2-dimensional arrival angle estimation with propagator method. IEEE Trans Antennas Propag 53:1622–1630
36. Xie Y, Xiong J, Li M, Jamieson K (2016) xD-track: leveraging multi-dimensional information for passive Wi-Fi tracking. Workshop on hot topics in wireless, pp 39–43
37. Tepedelenlioglu C, Abdi A, Giannakis G (2003) The Ricean K factor: estimation and performance analysis. IEEE Trans Wireless Commun 2(4):799–810
38. Han C, Wu K, Ni L (2014) WiFall: device-free fall detection by wireless networks. IEEE INFOCOM, 271–279
39. Chen Y, Wang W, Zhang Q (2014) Privacy-preserving location authentication in WiFi with fine-grained physical layer information. IEEE GLOBECOM, 4827–4832
40. Mcguire M, Plataniotis K (2003) Dynamic model-based filtering for mobile terminal location estimation. IEEE Trans Veh Technol 52(4):1012–1031
41. Yu K, Guo Y (2008) Improved positioning algorithms for nonline-of-sight environments. IEEE Trans Veh Technol 57(4):2342–2353
42. Guvenc I, Chong C, Watanabe F (2007) NLOS identification and mitigation for UWB localization systems. WCNC, 1571–1576
43. Guvenc I, Chong C, Watanabe F, Inamura H (2007) NLOS identification and weighted least-squares localization for UWB systems using multipath channel statistics. EURASIP J Adv Signal Process 2008(1):1–14
44. Zhou Z, Yang Z, Wu C, Sun W (2014) LiFi: line-of-sight identification with WiFi. INFOCOM, 2688–2696
45. Wu C, Yang Z, Zhou Z, Qian K (2015) PhaseU: real-time LOS identification with WiFi. INFOCOM, 2038–2046

46. Banerjee A, Dave R (2004) Validating clusters using the hopkins statistic. IEEE international conference on fuzzy systems, pp 25–29
47. http://standards.ieee.org/getieee802/download/802.11n-2009.pdf
48. Xie Y, Li Z, Li M (2015) Precise power delay profiling with commodity WiFi. MobiCom, 53–64
49. Li X, Zhang D, Xiong J, Wang Y (2016) Dynamic-MUSIC: accurate device-free indoor localization. UbiComp, 196–207
50. Scott D (2015) Multivariate density estimation: theory, practice, and visualization. Wiley
51. http://dhalperi.github.io/linux-80211n-csitool/index.html

Conclusions and Final Comments

Tariq S. Durrani, Wei Wang and Sheila M Forbes

This monographs is based on Chapters provided by experts from UK and China who had attendees an NSFC Newton Researcher Links Workshop held in Harbin in July 2017, and records the results of their latest research. With the objective of exploring emerging techniques in monitoring natural disasters, and associate mitigation activities, this monograph highlights a range of techniques for the purpose.

Each of the chapters covers the introduction of a specific monitoring technique, collection of data and the related analysis of the proposed technique to reflect performance in adverse circumstances. The First Chapter gives a broad introduction to the monograph. The Second Chapter explores the use of Ocean Bottom Pressure Gauges, attached to Ocean Bottom Seismometers, to establish a warning system for tsunamis, and specifically to observe the onset and occurrence of tsunamis, and offer new insights. This is a new approach to tsunami observations and compares favourably with the conventional technique of the DART (Deepwater Assessment and Reporting) approach; and offers two specific advantages—with a more dense installation than DARTs, they offer high spatial resolution coverage, and higher frequency resolution, due to the higher sampling rates, though they lead to a requirment of assessing a larger quantity of data. The authors have proposed new procedures for data collection and handling and then applied these to observations from two tsunamis to show the effectiveness of their approach.

T. S. Durrani (✉) · S. M. Forbes
University of Strathclyde, Glasgow, Scotland, UK
e-mail: durrani@strath.ac.uk

S. M. Forbes
e-mail: sheila.forbes@strath.ac.uk

W. Wang
Tianjin Normal University, Tianjin, People's Republic of China
e-mail: weiwang@tjnu.edu.cn

T. S. Durrani et al. (eds.), *Geological Disaster Monitoring Based on Sensor Networks*, Springer Natural Hazards, https://doi.org/10.1007/978-981-13-0992-2_12

The Third Chapter presented by a team mostly from the British Geological Survey, with partners in Florence, Italy; study the efficacy of satellite remote sensing in three diverse case of hazards—landslide inventory, tsunami induced damage, and geotechnical characterisation of a landslide in Italy.

In the case of the landslide inventory for the island of St. Lucia, the authors show the efficacy of Earth Observation Satellite data over land use surveys, and offer further insights into landscape response triggered by such hurricanes; and are vital in establishing relevant hazard and risk assessments. In the second case, the authors studied landslides in a small hamlet in Italy—Papanice, as a consequence of a severe rainfall. The lanslides were remotely monitored by means of multi-temporal InSAR, which provided ground displacement estimates at millimetre precision from TerraSAR-X images, integrated with other data to obtain a complete characterisation of the landslides. Interestingly, the integration of remote sensing and traditional geological/geotechnical investigations has suggested the installation of a superficial drainage system as the best intervention to mitigate the risk. The third case was a study of the most powerful Tsunami to affect Japan, in March 2011, and its impact near Sendai Airport. Using multi-temporal high-resolution satellite data, the researchers assessed the tsunami impact on the coast, as well as mapped the spatio-temporal trajectories of damage and reconstruction activities in the area. The work verified the value of the new (Earth Observation Satellites) spaceborne/airborne sensors and techniques developed in the last ten years, complemented by computational resources for running large-scale model simulations, leading to the development of 'best-practice' environmental information which enables planners to better manage geohazards.

The Fourth Chapter addressed the issue of rapid mapping of damages and damage assessment by proposing novel image segmentation and classification techniques. In particular the method combined advance image processing and pattern recognition techniques based on Adaptive Region Monitoring and Gravitational Self organising maps, to identify spectral textures and dynamically merging regions to classify cognate regions. These techniques create damage maps or detect and localize damage objects, delivering 'boundary-closed and spatial-continuous regions'. The algorithms were tested on real data from the Wenchuan Earthquake in China, and the work illustrated how inaccurate classification of regions can be avoided. The techniques are computing intensive and require significant segments of data, though the authors hold out the prospect of using 'deep learning' to overcome the above issues, and extend the work to multi-spectral and hyperspectral data.

The Fifth Chapter studied issues concerned with the deployment of Sensor Networks to monitor disasters and manage disaster regions. The authors conducted a review of current and emerging wireless communications technologies and associated IoT (Internet of Things) protocols to identify current approaches to Disaster Sensing and Monitoring. They posit that Wireless Sensor Networks, employing low energy sensor nodes capable of measuring and recording environmental conditions, exploiting 5G and emerging technologies, could be developed as intelligent and interconnected infrastructures to yield huge amounts of data, and combined with artificial intelligence would form the basis of the next generation of intelligent disaster management systems. The authors have studied the current IOT Standards LoRa/4G LTE and identified their limitations, and recommend that future work should focus on "developing robust fair data rate allocation and power control methods to address existing LoRa limitations and acquire optimised airtime, data rate, and energy consumption".

An important aspect of Disaster Management is the availability of tools which offer facilitation and assessment of risks and hazards. To this effect the Sixth Chapter is concerned with a new Earthquake Risk Assessment tool which is particularly applicable to regions with limited seismotectonic data—a not uncommon scenario. The team from the University of Sheffield has exploited a probabilistic framework to develop a methodology which utilises Monte Carlo techniques to simulate key hazard parameters such as magnitude, epicentre location, depth of hypocentre, and other geological parameters. Detailed analytical work was conducted to verify the approach, which is then tested against the real world scenario of the earthquake in the Turkish Region of Marmara. The results provided evidence to prove that the approach is acceptable for producing hazard maps. This is particularly useful for application in developing regions, for emergency response and mitigation purposes, where data is sparse or difficult to collect.

To observe, analyse, monitor, predict hazards and natural disasters, the previous chapters addressed wireless networks, satellite remote sensing, and underwater seismic sensor networks. The Seventh Chapter is a progression from the previous ones, and introduced the use of Unmanned Aerial Vehicles (UAVs), either as individual sources, for collecting information and for emergency communications; or in a universal network architecture to cover large swathes of areas. The UAVs are especially effective in disaster affected regions where the civilian infrastructure may have been demolished, roads blocked by debris etc. UAVs offer quick access to the disaster area and victims. Thus UAVs can be embedded into the entire life-cycle of disaster management.

In a very thorough piece of work, the authors proposed a universal network architecture for a UAV disaster management and studied issue related to the infrastructure-based connectivity between nodes, and associated characteristics, and considered adhoc and cooperative wireless based network. The proposed architectures were implemented on real time test-bed experiments between the two participating ground stations and a UAV, and the link quality for the performance of the network was studied; and network performance analysed in terms of bandwidth consumption, transmission latency and disruption prone networks. The Chapter provided significant pointers towards the Design considerations for a robust UAV based networked architecture.

The Eighth Chapter introduced the use of Radar Sensor Networks in the detection of humans, as in the debris of earthquakes and other disasters. The Chapter focused on the use of Ultra-Wideband (UWB) Radar systems, which are particularly effective in 'through the wall' penetration. Through elegant analysis, simulation and experimental set up, the author puts forward the efficacy of using UWB radars for detecting humans in disaster affected environments. The work identified six characteristics parameters, which are incorporated into fuzzy pattern recognition systems for robust recognition of human forms from the radar returns. The performance of the detection algorithms was analysed and it was shown that the discrimination is clear, though new tools need to be developed.

The Ninth Chapter was concerned with an important facet of natural disasters—that of the wind velocity. It is received wisdom that very high wind velocity is a precursor of disaster. Cyclones, rising flood waters, tsunamis are all accompanied by incipient increase in wind velocities. Extreme high winds are in themselves a danger to human life and cause damage to the environment. Thus accurate measurement of wind velocity is of immense importance. The Authors in this Chapter studied the use of non-invasive techniques for measuring wind velocity, and provided an analysis of the use of time-of-flight measurements to develop tomographic reconstruction of wind velocities.

This Chapter described the fundamentals of the simultaneous multi-channel time-of-flight measurements and their use in the tomographic reconstruction of 2D horizontal wind velocity distribution. The feasibility and effectiveness of the proposed methods were numerically validated in an extensive simulation study, which comprises an iterative process off line to establish the wind velocity fields, and results illustrated good quality reconstruction. The work needs to be extended to the reconstruction of wind velocity fields using real data.

The Tenth Chapter was concerned with resource optimisation when satellites are used in remote sensing of the environment during emergencies and in monitoring natural disasters. Taking the case of the Multi-beam Satellite Communication System, operating in S band, which plays a significant role in the provision of direct-to-user satellite mobile service; the authors developed an effective algorithm that covers joint power and bandwidth allocation in the presence of inter-beam interference, traffic demand, adverse channel conditions, capacity and demand, and environmental delays. This is a significant improvement on conventional algorithms that tend to allocate separately for power or bandwidth allocation. The work was reinforced by detailed simulations, which showed that the performance of the proposed approach is significantly superior in terms of throughput and energy efficiency, as compared to traditional integrated satellite-terrestrial spectrum sharing scheme at different user density; enhancing the integral energy efficiency at the cost of a small part of the throughput. As a follow on it would be good to test the performance with real data.

The Eleventh and final Chapter is devoted to the vexed issue of localisation accuracy of sources especially in an indoor environment. The concern arises when sources need to be identified in areas where remote sensing is not applicable, as within confined spaces and covered regions. The work reported in this Chapter combined analytical developments with an advanced experimental setup to verify theoretical predictions. Using a three antenna architecture, and commodity Wi-Fi devices, the authors were able to illustrate the detection capabilities that reflect sub-meter accuracy, without the need of any specialised hardware, thus opening the way to the use of commodity or off-the-shelf components. The merit of the approach resides in combining the Angle of Arrival (AoA) returns with Received

Signal Strength (RSS) measurements, to devise an algorithm that yields location with sub-meter accuracy, outperforming conventional systems relying only on AoA or RSS measurements. For detection purposes, an optimal threshold was set, based on the distribution of the AoA and Time of Flight, both for Line of Sight and Non-Line of Sight scenarios; the latter being a specially challenging case. The detection rate of 90% for the two scenarios is particularly impressive. Future work on extending the architecture to LTE and to 5G systems would be worthwhile.

Author Index

T. S. Durrani et al. (eds.), *Geological Disaster Monitoring Based on Sensor Networks*, Springer Natural Hazards, https://doi.org/10.1007/978-981-13-0992-2

Subject Index

T. S. Durrani et al. (eds.), *Geological Disaster Monitoring Based on Sensor Networks*, Springer Natural Hazards, https://doi.org/10.1007/978-981-13-0992-2

MIX
Papier aus verantwortungsvollen Quellen
Paper from responsible sources
FSC® C105338

If you have any concerns about our products,
you can contact us on
ProductSafety@springernature.com

In case Publisher is established outside the EU,
the EU authorized representative is:
Springer Nature Customer Service Center GmbH
Europaplatz 3, 69115 Heidelberg, Germany

Printed by Libri Plureos GmbH
in Hamburg, Germany